The Liar Paradox AND THE Towers of Hanoi

THE 10 GREATEST MATH PUZZLES OF ALL TIME

Marcel Danesi

John Wiley & Sons, Inc.

Published by John Wiley & Sons, Inc., Hoboken, New Jersey
Published simultaneously in Canada

Design and production by Navta Associates, Inc.

Library of Congress Cataloging-in-Publication Data:

Danesi, Marcel, date.
The liar paradox and the towers of Hanoi: the ten greatest math puzzles of all time / Marcel Danesi.
p. cm.
Includes bibliographical references and index.
ISBN 0-471-64816-7 (paper)
1. Mathematical recreations. I. Title.
QA95 .D29 2004
793.74—dc22

2003027191

Printed in the United States of America

10 9 8 7 6 5 4 3 2 1

To Alex and Sarah;
their existence is a gift and their life is a blessing.

CONTENTS

ACKNOWLEDGMENTS

I wish to thank the many people who have helped me, influenced me, and critiqued me over the years. First and foremost, I must thank all of the students I have had the privilege of teaching at the University of Toronto. They were a constant source of intellectual animation and enrichment. I must also thank Professor Frank Nuessel of the University of Louisville, for his unflagging help over many years. I am, of course, grateful to the editors at John Wiley for encouraging me to submit this manuscript to a publishing house that is renowned for its interest in mathematics education. It is my second book for Wiley. I am particularly grateful to Stephen Power, Jeff Golick, and Michael Thompson for their expert advice, and to Kimberly Monroe-Hill and Patricia Waldygo for superbly editing my manuscript, greatly enhancing its readability. Needless to say, any infelicities that this book may contain are my sole responsibility.

Finally, a heartfelt thanks goes out to my family, which includes Lucy (my wife), Alexander and Sarah (my grandchildren), Danila (my daughter), Chris (my son-in-law), and Danilo (my father), for the patience they have shown me during the research and the writing of this book. I truly must beg their forgiveness for my having been so cantankerous and heedless of family duties.

Introduction

PUZZLES ARE AS OLD AS HUMAN HISTORY. They are found in cultures throughout the ages. Why is this so? What are puzzles? What do they reveal about the human mind? Do they have any implications for the study of mathematics?

This book attempts to answer some of these questions. Its main focus is on showing how certain ideas in mathematics originated in the form of puzzles. I use the word **puzzle** in its basic sense, to mean a challenging problem that conceals a nonobvious answer. I do not use it in the figurative sense of "anything that remains unsolved," even though the two meanings share a lot of semantic territory, as the mathematician Keith Devlin recently demonstrated in his fascinating book on the seven greatest unsolved mathematical puzzles of our time (*The Millennium Problems*, Basic Books, 2002).

In the humanities and the arts, there is a long-standing tradition of identifying the masterpieces—the great novels, the great symphonies, and the great paintings—as the most illuminating things to study. Books are written and courses taught on them. Mathematics, too, has its "great" problems. Significantly, most of these were originally devised as clever puzzles. So, in line with teaching practices in literature, music, and the fine arts, this book introduces basic mathematical ideas through ten puzzle masterpieces. Needless to say, so many ingenious puzzles have been invented throughout history that it would be brazenly presumptuous to claim that I chose the ten best. In reality, I went on a mathematical dig to unearth ten puzzles that were demonstrably pivotal in shaping mathematical history and that, I believe, most mathematicians would also identify as among the most important ever devised.

The Uses of This Book

Above all else, this book can be read to gain a basic understanding of what puzzles are all about and to grasp their importance to mathematics. Anyone wishing to acquire a basic skill at puzzle-solving and at doing elementary mathematics can also use it profitably as a self-study manual. It is not meant, however, to be a collection of puzzles, challenging or otherwise. There are many such books on the market. Rather, it is a manual on the relationship between puzzles and mathematics. In a word, it is written for "beginners," not for inveterate puzzle solvers.

Teachers will find that as a classroom text, it covers the same kind of topics found in more traditional introductory math texts, even though it takes a different, more creative slant toward them. Students can discuss each puzzle and its implications for the study of mathematics, then can follow up on the sources in the Further Reading sections. They can also come up with their own puzzle activities or research each great puzzle further and report their findings to the class.

This book is based on materials I prepared for a noncredit course that I've taught at the University of Toronto for over a decade. The course is aimed at so-called math phobics. I have consistently found that an engagement with puzzles allows such students to gain confidence and go on to more complex areas of mathematics with little or no difficulty. The congratulatory e-mails that I receive from ex-students are a source of great pride. Nothing makes teachers happier than to witness students become proficient at what they teach! I truly hope that this book will allow readers to achieve similar results. I invite the readers of this book to contact me at my e-mail address any time they wish: marcel.danesi@utoronto.ca.

Format

Each of the ten chapters is divided into five sections: The Puzzle, Mathematical Annotations, Reflections, Explorations, and Further Reading.

The Puzzle

Each puzzle is explained in an easy-to-follow manner. Complete adherence to the original solutions and to the mathematical implications that ensued from them would make some of the puzzles extremely difficult to understand. In such cases, I made appropriate modifications. Nevertheless, I tried to retain the spirit of each puzzle and its solution. Regarding the reader's background knowledge, I took very little for granted. Every mathematical symbol, notation, formula, and concept introduced into the discussion of a puzzle is fully explained. For example, if knowledge of exponents is required at some point, then I provide a brief explanatory note on that topic in a sidebar.

In-depth discussions of the selected puzzles can be found in W. W. Rouse Ball's *Mathematical Recreations and Essays*, which was published in 1892 and was since reissued in many more editions. Readers can also refer to the writings of Martin Gardner (1914–) and Raymond Smullyan (1919–), if they would like more exposure to puzzle-solving and are interested in complementary treatments of the relationship between puzzles and mathematics and logic. Their writings are listed in some of the Further Reading sections at the end of the chapters. For thirty years, starting in 1956, Gardner wrote a famous puzzle column for *Scientific American*. Smullyan has written a series of ingenious puzzle books designed to strip down logical reasoning to its essentials. There are also magazines and journals, such as the *Journal of Recreational Mathematics*, *Eureka*, and *Games*, that readers can consult to extend their involvement with **puzzlemath**. However, I warn the neophyte puzzlist that it would be a difficult task indeed to directly tackle the subject matter of these sources, without some elementary training beforehand. I hope this book will provide exactly that.

Mathematical Annotations

The discussion of each puzzle is followed by annotations on its implications for a specific area of mathematics or for mathematics generally. Every notion introduced in the discussion is explained fully. Even common concepts, such as *prime number* and *composite number*, are clarified when they are introduced. A glossary of such terms is provided at the back, for the reader's convenience. The only assumption I make is that the reader knows how to carry out basic arithmetical operations such as addition, subtraction, multiplication, and division, and knows generally what an equation is. In more detail, an **equation** is a statement asserting that two expressions are equal or the same. It is usually written as a line of symbols that are separated into left and right sides and joined by an equal sign (=). For example, in $x + 5 = 8$, the expression $(x + 5)$ is the left side of the equation and the number 8 is the right side. The left side of this equation will be equal to the right side when the letter x is replaced by the number 3—$(3 + 5 = 8)$. I have taken virtually nothing else for granted.

Reflections

After the mathematical annotations, I have added my own reflections on the puzzle or its mathematical implications.

Explorations

This section provides follow-up exploratory exercises that allow readers to engage directly in puzzle-solving. Answers and detailed explanations to the exercises are found at the back of the book. A word of advice is in order. Do not be discouraged if, at first, you have difficulty with a specific exercise. Try your best to solve it before you read the explanation at the back. This will allow you to grasp the spirit of the puzzle.

The explorations are numbered consecutively across chapters. This allows readers who might prefer to use the book primarily for its puzzle-solving value to go to the exercises directly in sequence. There are eighty-five brainteasers—no less than are usually found in most puzzle books available on the market.

Further Reading

A list of the sources I used is provided at the end of each chapter. Readers who are interested in expanding their knowledge of a certain puzzle or a related area of mathematics can consult the sources directly.

The Riddle of the Sphinx

Let us consider that we are all partially insane.
It will explain us to each other; it will unriddle many
riddles; it will make clear and simple many things
which are involved in haunting and harassing
difficulties and obscurities now.

MARK TWAIN (1835–1910)

IF WE VISIT THE CITY OF GIZA in Egypt today, we cannot help but be overwhelmed by the massive sculpture known as the Great Sphinx, a creature with the head and the breasts of a woman, the body of a lion, the tail of a serpent, and the wings of a bird. Dating from before 2500 B.C., the Great Sphinx magnificently stretches 240 feet (73 meters) in length and rises about 66 feet (20 meters) above us. The width of its face measures an astounding 13 feet, 8 inches (4.17 meters).

Legend has it that a similarly enormous sphinx guarded the entrance to the ancient city of Thebes. The first recorded puzzle in human history comes out of that very legend. The Riddle of the Sphinx, as it came to be known, constitutes not only the point of departure for this book but the starting point for any study of the relationship between puzzles and mathematical ideas. As humankind's earliest puzzle, it is among the ten greatest of all time. Riddles are so common, we hardly ever reflect upon what they are. Their appeal is ageless and timeless. When children are posed a simple riddle, such as "Why did the chicken cross the road?" without any hesitation whatsoever, they seek an answer to it, as if impelled by some unconscious mythic instinct to do so.

Readers may wonder what a riddle shrouded in mythic lore has to do with mathematics. The answer to this will become obvious as they work their way through this chapter. Simply put, in its basic structure, the Riddle of the Sphinx is a model of how so-called insight thinking unfolds. And this form of thinking undergirds all mathematical discoveries.

The Puzzle

According to legend, when Oedipus approached the city of Thebes, he encountered a gigantic sphinx guarding the entrance to the city. The menacing beast confronted the mythic hero and posed the following riddle to him, warning that if he failed to answer it correctly, he would die instantly at the sphinx's hands:

> What has four feet in the morning, two at noon, and three at night?

THE OEDIPUS LEGEND

In Greek mythology, the oracle (prophet) at Delphi warned King Laius of Thebes that a son born to his wife, Queen Jocasta, would grow up to kill him. So, after Jocasta gave birth to a son, Laius ordered the baby taken to a mountain and left there to die. As fate would have it, a shepherd rescued the child and brought him to King Polybus of Corinth, who adopted the boy and named him Oedipus.

Oedipus learned about the ominous prophecy during his youth. Believing that Polybus was his real father, he fled to Thebes, of all places, to avoid the prophecy. On the road, he quarreled with a strange man and ended up killing him. At the entrance to Thebes, Oedipus was stopped by an enormous sphinx that vowed to kill him if he could not solve its riddle. Oedipus solved it. As a consequence, the sphinx took its own life. For ridding them of the monster, the Thebans asked Oedipus to be their king. He accepted and married Jocasta, the widowed queen.

Several years later, a plague struck Thebes. The oracle said that the plague would end when King Laius's murderer had been driven from Thebes. Oedipus investigated the murder, discovering that Laius was the man he had killed on his way to Thebes. To his horror, he learned that Laius was his real father and Jocasta his mother. In despair, Oedipus blinded himself. Jocasta hanged herself. Oedipus was then banished from Thebes. The prophecy pronounced at Delphi had come true.

The fearless Oedipus answered, "Humans, who crawl on all fours as babies, then walk on two legs as grown-ups, and finally need a cane in old age to get around." Upon hearing the correct answer, the astonished sphinx killed itself, and Oedipus entered Thebes as a hero for ridding the city of the terrible monster that had kept it captive for so long.

Various versions of the riddle exist. The previous one is adapted from the play *Oedipus Rex* by the Greek dramatist Sophocles (c. 496–406 B.C.). Following is another common statement of the riddle, also dating back to antiquity:

> What is it that has one voice and yet becomes four-footed, then two-footed, and finally three-footed?

Whatever its version, the Riddle of the Sphinx is the prototype for all riddles (and puzzles, for that matter). It is intentionally constructed to harbor a nonobvious answer—namely, that life's three phases of infancy, adulthood, and old age are comparable, respectively, to the three phases of a day (morning, noon, and night). Its function in the Oedipus story, moreover, suggests that puzzles may have originated as tests of intelligence and thus as probes of human mentality. The biblical story of Samson is further proof of this. At his wedding feast, Samson, obviously wanting to impress the relatives of his wife-to-be, posed the following riddle to his Philistine guests (Judges 14:14):

> Out of the eater came forth meat and out of the strong came forth sweetness.

He gave the Philistines seven days to come up with the answer, convinced that they were incapable of solving it. Samson contrived his riddle to describe something that he once witnessed—a swarm of bees that made honey in a lion's carcass. Hence, the wording of the riddle: the "eater" = "swarm of bees"; "the strong" = the "lion"; and "came forth sweetness" = "made honey." The deceitful guests, however, took advantage of the seven days to coerce the answer from Samson's wife. When they gave Samson the correct response, the mighty biblical hero became enraged and declared war against all Philistines. The ensuing conflict eventually led to his own destruction.

The ancients saw riddles as tests of intelligence and thus as a means through which they could gain knowledge. This explains why the Greek priests and priestesses (called oracles) expressed their prophecies in the form of riddles. The implicit idea was, evidently, that only people who could penetrate the language of the message would unravel its concealed prophecy.

However, not all riddles were devised to test the acumen of mythic heroes. The biblical kings Solomon and Hiram, for example, organized riddle contests simply for the pleasure of outwitting each other. The ancient Romans made riddling a recreational activity of the Saturnalia, a religious event that they celebrated from December 17 to 23. By the fourth century A.D., riddles had, in fact, become so popular for their "recreational value" that the memory of their mythic origin started to fade. In the tenth century, Arabic scholars used riddles for pedagogical reasons—namely, to train students of the law to detect linguistic ambiguities. This coincided with the establishment of the first law schools in Europe.

Shortly after the invention of the printing press in the fifteenth century, some of the first books ever printed for popular entertainment were collections of riddles. One of these, titled *The Merry Book of Riddles,* was published in 1575. Here is a riddle from that work:

> He went to the wood and caught it,
> He sate him downe and sought it;
> Because he could not finde it,
> Home with him he brought it.
>
> (answer: a thorn caught on a foot)

By the eighteenth century, riddles were regularly included in many newspapers and periodicals. Writers and scholars often composed riddles. The American inventor Benjamin Franklin (1706–1790), for instance, devised riddles under the pen name of Richard Saunders. He included them in his *Poor Richard's Almanack,* first published in 1732. The almanac became an unexpected success, due in large part to the popularity of its riddle section. In France, no less a literary figure than the great satirist Voltaire (1694–1778) penned brain-teasing riddles, such as the following one:

> What of all things in the world is the longest, the shortest, the swiftest, the slowest, the most divisible and most extended, most regretted, most neglected, without which nothing can be done, and with which many do nothing, which destroys all that is little and ennobles all that is great?
>
> (answer: time)

The ever-increasing popularity of riddles in the nineteenth century brought about a demand for more variety. This led to the invention of a new riddle genre, known as the *charade.* Charades are solved one syllable or line at a time, by unraveling the meanings suggested by separate syllables, words, or lines. In the nineteenth century, this led to the *mime charade*, which became, and continues to be, a highly popular game at social gatherings. It is played by members of separate teams who act out the meanings of various syllables of a word, an entire word, or a phrase in pantomime. If the answer to the charade is, for example, "baseball," the syllables *base* and *ball* of that word are the ones normally pantomimed. By the end of the century, riddles were firmly embedded in European and American recreational culture and remain so to this very day.

Mathematical Annotations

The question that the legendary sphinx asked Oedipus seems to defy an answer at first. What bizarre creature could possibly have four, then two, and finally three legs, in that order? Wresting an answer from the riddle requires us to think imaginatively, not linearly. This very type of imaginative thinking undergirds all true mathematical inquiry.

Problem-Solving

PROBLEM-SOLVING METHODS AND STRATEGIES

Deduction: This involves applying previous knowledge to the problem.

Induction: This involves reasoning from particular facts given in the problem, to reach a general conclusion.

Insight thinking: This involves making guesses or following up on hunches that come from trial-and-error approaches to the problem.

Riddles highlight how puzzles differ in general from typical mathematical problems, such as those found in school textbooks. The latter are designed to help students do something systematically (for example, add large numbers, solve equations, prove theorems, etc.). To grasp the difference, consider two typical textbook problems. Here's the first one:

> Prove that the vertically opposite angles formed when two straight lines intersect are equal.

The method used to solve this type of problem is called **deduction**. It involves applying previous knowledge to the problem at hand.

Start by drawing a diagram that shows all the relevant features of the problem. The two straight lines can be labeled **AB** and **CD**, and two of the four vertically opposite angles formed by their intersection can be labeled x and y. One of the angles between x and y can be labeled z, as shown:

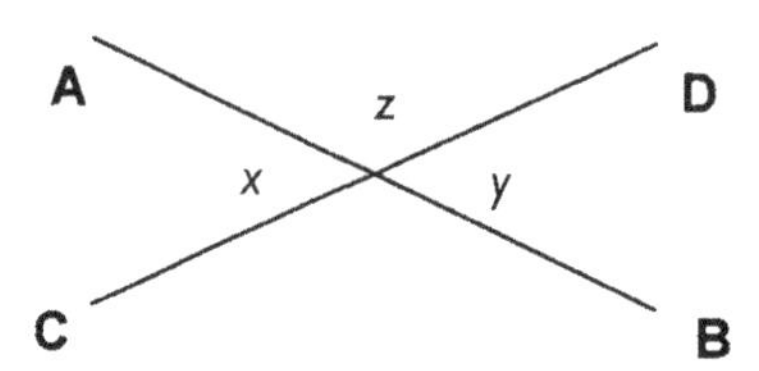

The problem asks us, in effect, to prove that x and y (being vertically opposite angles) are equal. There are, of course, two other vertically opposite angles formed by the intersection of the two lines, but they need not be considered here because the method of proof and the end result are the same. The proof hinges on previous knowledge—specifically, that a straight line is an angle of 180 degrees. Consider **CD** first. As a straight line, it is (as mentioned) an angle of 180 degrees. Now, notice that **CD** is composed of two smaller angles on the diagram, x and z. So, logically, these two must add up to 180 degrees—a statement that can be represented with the equation $x + z = 180°$. The equation reads as follows: "Angle x and angle z when added together equal 180°."

Now, consider **AB**. Notice that it, too, is composed of two smaller angles on the diagram, y and z. These two angles must also add up to 180 degrees—a fact that can be similarly represented with an equation: $y + z = 180°$. The two equations just discussed are listed as follows:

1. $x + z = 180°$
2. $y + z = 180°$

They can be rewritten as follows:

3. $x = 180° - z$
4. $y = 180° - z$

If you have forgotten your high school algebra, the reason we can do this is that whatever is done to one side of an equation must also be done to the other. Think of the two sides of an equation as the two pans on a balancing scale, with equal weights in each pan. The weights are analogous to the expressions on either side of an equation. If we want to maintain balance, any weight we take from one of the pans (such as the left one) we must also take from the other (the right one). In like fashion, if we subtract z from the left side of equation 1, we must also subtract it from its right side. The result is equation 3, which shows that z has been subtracted from both sides. Note that when z is subtracted from itself on the left side ($z - z$), it leaves 0—a result that is not normally indicated. Subtracting z from both sides of equation 2 yields equation 4.

Now, since two things that are equal to the same thing are equal to each other (for example, if Alex is six feet tall and Sarah is six feet tall, then the two people are equal in height), we can deduce that $x = y$, since equation 3 shows that x is equal to $(180° - z)$, and equation 4 shows that y is equal to the same expression $(180° - z)$. It is not necessary to figure out what the value of the expression is. Whatever it is, the fact remains that both x and y will be equal to it. We can now conclude that "any two vertically opposite angles produced by the intersection of two straight lines are equal," because we did not assign a specific value to either angle. When a proof is generalizable in this way, it is called a **theorem**.

POLYGONS

A polygon is a closed plane (two-dimensional) figure. Examples of polygons are triangles, quadrilaterals such as rectangles and squares, pentagons (five-sided figures), and hexagons (six-sided figures).

The sum of the three angles in any triangle is 180°, no matter what type of triangle it is (see chapter 5).

Here's our second textbook problem:

Develop a formula for the number of degrees in any polygon.

Solving this problem entails a different kind of strategy, known as **induction**. This involves extracting a generalization on the basis of observed facts. Consider a triangle first—the **polygon** with the least number of sides. The sum of the angles in a triangle is 180 degrees (see chapter 5 for the relevant proof).

Next, consider any quadrilateral (a four-sided figure). **ABCD** is one such figure:

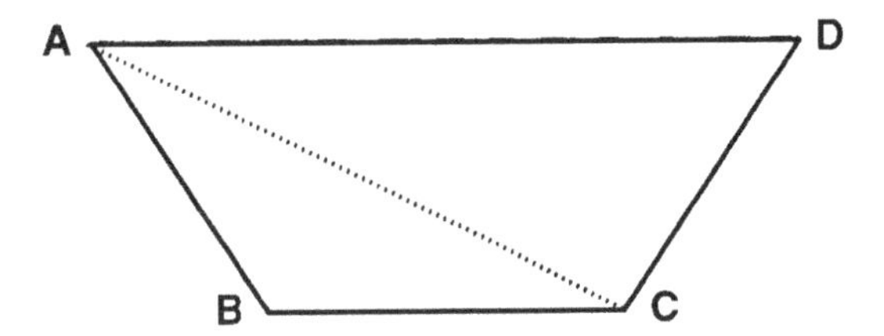

Notice that this figure can be divided into two triangles as shown (triangle **ABC** and triangle **ADC**). By doing this, we have discovered that the sum of the angles in the quadrilateral is equivalent to the sum of the angles in two triangles, namely, 180° + 180° = 360°.

Next, consider the case of a pentagon (a five-sided figure). **ABCDE**, as follows, is one such figure:

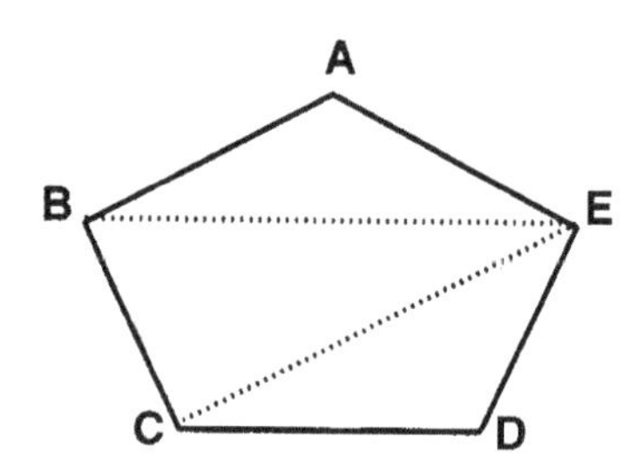

Since the pentagon can be divided into three triangles, as shown (triangle **ABE**, triangle **BEC**, and triangle **ECD**), we have again discovered a simple fact—namely, that the sum of its angles is equivalent to the sum of the angles in three triangles: 180° + 180° + 180° = 540°.

Continuing in this way, we can show just as easily that the number of angles in a hexagon (a six-sided figure) is equal to the sum of the angles in four triangles; in a heptagon (a seven-sided figure), to the sum of the angles in five triangles; and so on. Let's now attempt to generalize what we have apparently discovered. The letter n can be used to represent any number of sides, and the term *n-gon* can be used to refer to any polygon—that is, to a polygon with an unspecified number of sides. The previous observations suggest that the number of triangles that can be drawn in any polygon is "two less" than the number of sides that make up the polygon. For example, in a quadrilateral, we can draw two triangles, which is "two less" than the number of its sides (4), or (4 – 2); in a pentagon, we can draw three triangles, which is, again, "two less" than the number of its sides (5), or (5 – 2); and so on. In the case of a triangle, this rule also applies, since we can draw in it one and only one triangle (itself). This also is "two less" than the number of its sides (3), or (3 – 2). In an *n-gon*, therefore, we can draw $(n - 2)$ triangles. To summarize:

TABLE 1-1: CALCULATING THE TRIANGLES IN A POLYGON

Number of Sides in the Polygon	Number of Triangles That Can Be Drawn in the Polygon
3 (= triangle)	(3 – 2) = 1 triangle
4 (= quadrilateral)	(4 – 2) = 2 triangles

Number of Sides in the Polygon	Number of Triangles That Can Be Drawn in the Polygon
5 (= pentagon)	$(5-2) = 3$ triangles
6 (= hexagon)	$(6-2) = 4$ triangles
7 (= heptagon)	$(7-2) = 5$ triangles
. . .	. . .
n (= n-gon)	$(n-2)$ triangles

Since we know that there are 180 degrees in a triangle, then there will be $(4-2)\ 180°$ in a quadrilateral, $(5-2)\ 180°$ in a pentagon, and so on. Thus, in an *n-gon*, there will be $(n-2)\ 180°$:

TABLE 1-2: DETERMINING THE DEGREES IN A POLYGON

Number of Sides in the Polygon	Number of Triangles That Can Be Drawn in the Polygon	Sum of Degrees of the Angles in the Polygon
3	$(3-2) = 1$	$180° \times 1 = 180°$
4	$(4-2) = 2$	$180° \times 2 = 360°$
5	$(5-2) = 3$	$180° \times 3 = 540°$
6	$(6-2) = 4$	$180° \times 4 = 720°$
7	$(7-2) = 5$	$180° \times 5 = 900°$
. . .	. . .	. . .
n	$(n-2)$	$180° \times (n-2) = 180°\ (n-2)$

The formula can be written as

$$(n-2)\ 180°$$

or as

$$180°\ (n-2).$$

Now we can determine the number of degrees in any polygon in a straightforward fashion. For example, in the case of an octagon, $n = 8$. Plugging this value into our formula will yield the number of degrees in an octagon:

$$(n-2)\ 180° = (8-2)\ 180° = 6 \times 180° = 1{,}080°.$$

The thing to note about this problem's solution is that it involves generalizing from particular instances. That is the sum and the substance of

COMMUTATIVITY

Changing the order of the factors (numbers) in a multiplication does not change the result (the product). This property of multiplication is known as **commutativity**. Examples include:

$$2 \times 3 = 3 \times 2 = 6$$

$$4 \times 9 = 9 \times 4 = 36$$

In general (n = any number, m = any other number),

$$n \times m = m \times n.$$

It can also be written as

$$nm = mn.$$

So, applying the principle of commutativity to our case, we get

$$180° (n - 2) = (n - 2)\ 180°.$$

This same property, incidentally, holds for addition. Examples are:

$$2 + 3 = 3 + 2 = 5$$

$$4 + 9 = 9 + 4 = 13$$

In general,

$$n + m = m + n.$$

Commutativity does not hold for either subtraction or division, as you can see for yourself ($\neq$ stands for "does not equal"). Some examples are:

$$7 - 4 \neq 4 - 7$$

$$9 \div 3 \neq 3 \div 9$$

inductive reasoning. However, a caveat is in order with respect to such reasoning. Consider the following arithmetical computations—multiplications are on the left and additions are on the right:

Multiplication	**=**	**Addition?**
$2 \times 2 = 4$		$2 + 2 = 4$
$\frac{3}{2} \times 3 = 4\frac{1}{2}$		$\frac{3}{2} + 3 = 4\frac{1}{2}$
$\frac{4}{3} \times 4 = 5\frac{1}{3}$		$\frac{4}{3} + 4 = 5\frac{1}{3}$
$\frac{5}{4} \times 5 = 6\frac{1}{4}$		$\frac{5}{4} + 5 = 6\frac{1}{4}$

From these examples, we might conclude that multiplying numbers always produces the same result as adding them. But, of course, that is not true. Therefore, certain conditions apply when using the method of induction to solve problems. We will return to this topic in chapter 5.

Insight Thinking

What distinguishes the Riddle of the Sphinx from problems such as those we just solved is that the solution strategy is not as predictable. Solving riddles requires **insight thinking**. This can be characterized, essentially, as the act or the outcome of intuitively grasping the inward or hidden nature of a problem. Humanity's first puzzle is a model of how insight thinking unfolds.

The relevant insight required to solve the Riddle of the Sphinx is not to interpret its words literally but to do so metaphorically. Most riddles are based on the various meanings of a word. Consider the following example:

> What has four wheels and flies?
>
> (answer: a garbage truck)

The answer makes sense only when we realize that the word *flies* has two meanings—as a verb ("to move through the air") and as a noun ("an insect with two wings"). A garbage truck is indeed something that has "four wheels" and "flies" that surround it, given that flies are attracted to garbage.

It might be instructive to turn the tables around and create a riddle ourselves. Take, for example, the word *smile*. In English, a smile is said to be something that, like clothing, can be worn. This is why we speak of "wearing a smile," "taking a smile off one's face," and so on. Now, we propitiously can use this very linguistic convention to phrase our riddle:

> I am neither clothes nor shoes, yet I can be worn and taken off when not needed any longer. What am I?

Parenthetically, riddles can also be composed to provide humor. Take, for example, the classic children's riddle "Why did the chicken cross the road?" The number of replies to this question is infinite. Here are three possible answers:

1. To get to the other side.
2. Because it was taken across by a farmer.
3. Because a fox was chasing it.

All three answers tend to evoke moderate laughter, similar to the kind that the punch line of a joke would elicit. Riddles of this kind abound, revealing that they have a lot in common with humor.

Insight thinking is the defining characteristic of how most (if not all) puzzles are solved. As an example, consider the following classic puzzle:

> Without letting your pencil leave the paper, can you draw four straight lines through the following nine dots?

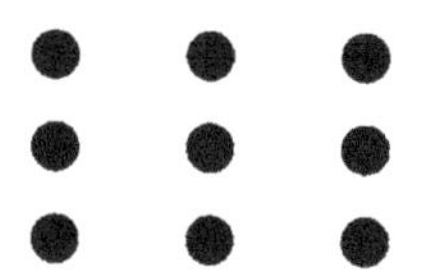

At first, people tend to approach this puzzle by joining up the dots as if they were located on the perimeter (boundary) of an imaginary square or a flattened box:

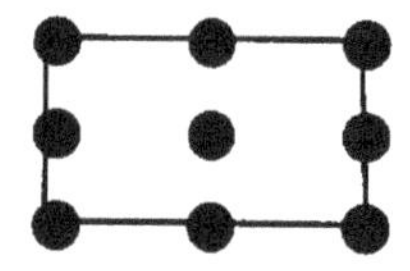

But this reading of the puzzle does not yield a solution, no matter how many times one tries to draw four straight lines without lifting the pencil. A dot is always "left over," as the following three attempts show:

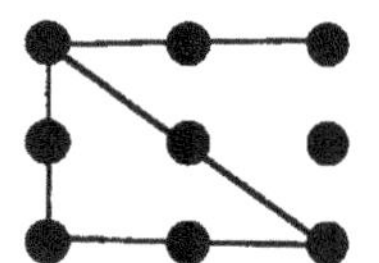
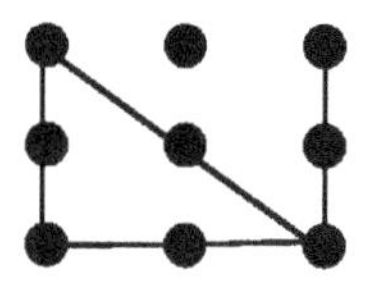
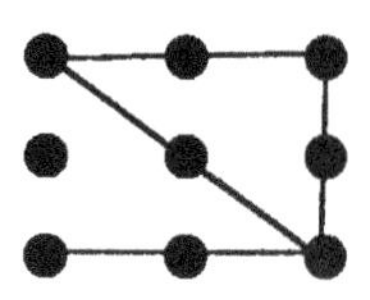

At this point, intuition comes into play: "What would happen if I extend one or more of the four lines beyond the box?" That hunch turns out, in fact, to be the relevant insight.

Start by putting the pencil on, say, the bottom left dot, tracing a straight line upward through the two dots above it and stopping at a point "outside the box," when you can see that it is in line diagonally with the two dots below it. You could start with any of the four corner dots and produce a solution (as you may wish to confirm yourself):

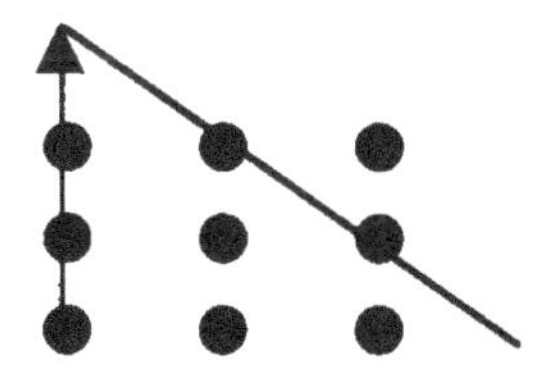

Second, trace a straight line diagonally downward through the two dots. Stop when you see that your second line is in horizontal alignment with the three bottom dots:

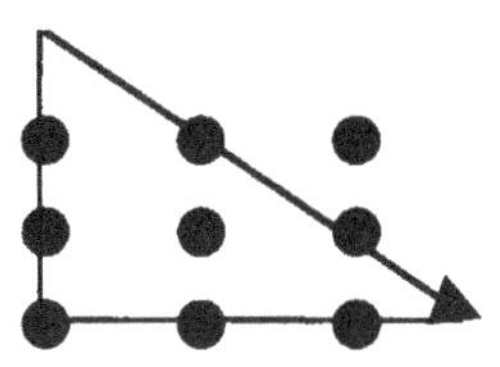

Draw your third line through the bottom dots:

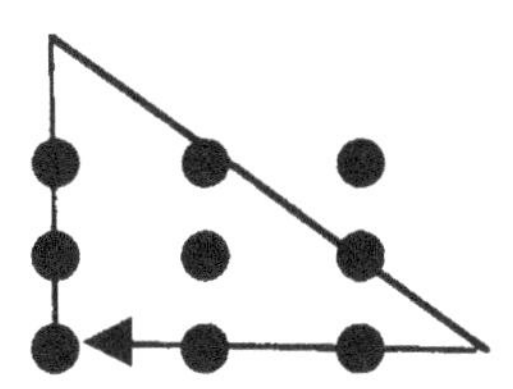

Finally, draw your fourth line through the remaining dots:

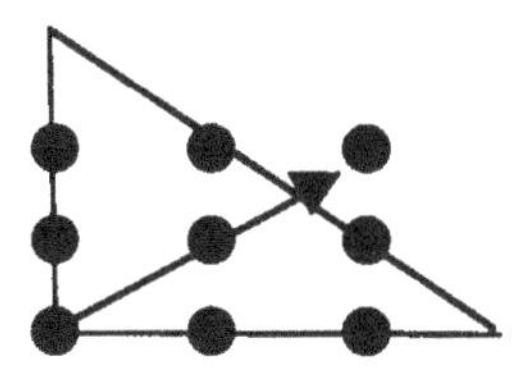

Incidentally, this puzzle is the probable source of the common expression "thinking outside the box." The reason for this is self-explanatory.

Solving puzzles may, at times, involve the use of other forms of thinking. But it is the intuitive trial-and-error form that dominates. The word *puzzle,* incidentally, comes from the Middle English word *poselen,* "to bewilder, confuse." It is an apt term because, unlike the typical problems found in mathematics textbooks, puzzles at first generate bewilderment and confusion, at the same time that they challenge our wits. As Helene

Hovanec has stated in her delectable book *The Puzzler's Paradise* (see Further Reading), the lure of puzzles lies in the fact that they "simultaneously conceal the answers yet cry out to be solved," piquing solvers to pit "their own ingenuity against that of the constructors."

Consider one more classic puzzle, devised by the French Jesuit poet and scholar Claude-Gaspar Bachet de Mézirac (1581–1638)—a puzzle that he included in his 1612 collection titled *Problèmes plaisans et délectables qui se font par les nombres* ("Amusing and Delightful Number Problems"):

> What is the least number of weights that can be used on a scale to weigh any whole number of pounds of sugar from 1 to 40 inclusive, if the weights can be placed on either of the scale pans?

We might, at first, be tempted to conclude that six weights of 1, 2, 4, 8, 16, and 32 pounds would do the trick. The reasoning would go somewhat as follows. We could weigh 1 pound of sugar by putting the 1-pound weight on the left pan, pouring sugar into the right pan until both pans balance. We could weigh 2 pounds of sugar by putting the 2-pound weight on the left pan, pouring sugar on the right pan until the pans balance. We could weigh 3 pounds of sugar by putting the 1-pound and the 2-pound weights on the left pan, pouring sugar on the right pan until the pans balance. And so on, and so forth. In this way, we could weigh any number of integral (whole-number) pounds of sugar from 1 pound to 40 pounds.

EXPONENTS

An *exponent* (also called a *power*) is a superscript digit or letter attached to the right of a number, indicating how many times the number is to be multiplied by itself. For example, in 3^4 the superscript digit 4 indicates that the number 3 is to be multiplied by itself four times:

$$3^4 = 3 \times 3 \times 3 \times 3.$$

The term 3^4 is read: "3 to the power of four" or "3 to the fourth power."

Exponential representation is shorthand form for repeated multiplication. Examples include:

$$2^1 = 2$$
$$3^2 = 3 \times 3$$
$$5^3 = 5 \times 5 \times 5$$
$$\ldots$$
$$n^4 = n \times n \times n \times n$$

(continued)

Any number to the zero power is always 1, no matter what the number is (see chapter 6). Examples include:

$3^0 = 1$
$9^0 = 1$
...
$n^0 = 1$

However, since the puzzle allows us to put the weights on both pans of the scale, the weighing can be done—Aha!—with only four weights of 1, 3, 9, and 27 pounds. The reason for this is remarkably simple—placing a weight on the right pan, along with the sugar, is equivalent to taking its weight away from the total weight on the left pan. Think about this for a moment. For example, if 2 pounds of sugar are to be weighed, we would put the 3-pound weight on the left pan and the 1-pound weight on the right pan. The result is that there are 2 pounds less on the right pan. We will therefore get a balance when we pour the missing 2 pounds of sugar on the right pan.

The four weights are, upon closer scrutiny, powers of 3:

$1 = 3^0$

$3 = 3^1$

$9 = 3^2$

$27 = 3^3$

The choice of these weights works because each of the whole numbers from 1 to 40 (= the required weights) turns out to be either a multiple or a power of 3, or else one more or less than a multiple or a power of 3. Thus, each of the first forty integers can be expressed with the first four powers of 3:

$1 = 3^0$ $(= 1)$

$2 = 3^1 - 3^0$ $(= 3 - 1)$

$3 = 3^1$ $(= 3)$

$4 = 3^1 + 3^0$ $(= 3 + 1)$

$5 = 3^2 - 3^1 - 3^0 = 3^2 - (3^1 + 3^0)$ $(= 9 - 3 - 1 = 6 - 1)$

... ...

$40 = 3^3 + 3^2 + 3^1 + 3^0$ $(= 27 + 9 + 3 + 1 = 39 + 1)$

Since the four powers of 3 represent our weights, all we have to do is "translate" addition in the previous layout as the action of putting weights on the left pan and subtraction as the action of putting weights on the right pan (along with the sugar). The following chart gives an indication of how this can be done. Readers may wish to complete it on their own:

TABLE 1-3: MÉZIRAC'S WEIGHT PUZZLE

Amount of Sugar to Be Weighed	Weight to Be Placed on the Left Pan	Weight Added to the Right Pan along with the Sugar
1	3^0 (= 1)	None
2	3^1 (= 3)	3^0 (= 1)
3	3^1 (= 3)	None
4	$3^1 + 3^0$ (= 3 + 1)	None
5	3^2 (= 9)	$3^1 + 3^0$ (= 4)
...	...	...
40	$3^3 + 3^2 + 3^1 + 3^0$ (= 27 + 9 + 3 + 1)	None

Reflections

The Riddle of the Sphinx is the first example in history of a true puzzle. Its origin in myth resonates to this day in stories composed for children. The heroes in such stories typically face challenges that are designed to test not only their physical mettle but also their mental ability to solve riddles. As such narrative traditions suggest, we perceive riddles as "miniature revelations" of truth. What are philosophy and science, after all, if not attempts to answer the riddles that life poses?

Mathematical inquiry, too, seems to be guided by an inborn need to model perplexing ideas in the form of puzzles. This is perhaps why some of the greatest questions of mathematical history were originally framed as puzzles. Solving them required a large dose of insight thinking. In many cases, the insight took centuries and even millennia to come to fruition. But, eventually, it did, leading to significant progress in mathematics. It would seem that in order to enter the "Thebes" of mathematical knowledge, we must first solve challenging riddles, not unlike the Riddle of the Sphinx.

Explorations

Riddles

1. What can be thrown away when it is caught but must be kept when it is not caught?

2. What possible creature is unlike its mother and does not resemble its father? You should also know that it is of mingled race and incapable of producing its own progeny.

3. I scare away my master's foes by bearing weapons in my jaws, yet I flee before the lashings of a little child. What am I?

4. It is something red, blue, purple, and green. Everyone can easily see it, yet no one can touch it or even reach it. What is it?

5. Before my birth I had a name, but it changed the instant I was born. And when I am no more, I will be called by my father's name. In sum, I change my name three days in a row, yet I live but one day. Who or what am I?

6. What belongs to you, which others use more than you do?

7. Create riddles based on the following words:

A. justice

B. friendship

C. love

D. time

Deductive Reasoning

8. A triangle, **ABC**, is inscribed in a semicircle ("half circle"), with its base, **BC**, resting on the diameter. Prove that the angle opposite the base, ∠**BAC**, is equal to 90 degrees. The sign ∠ stands for "angle":

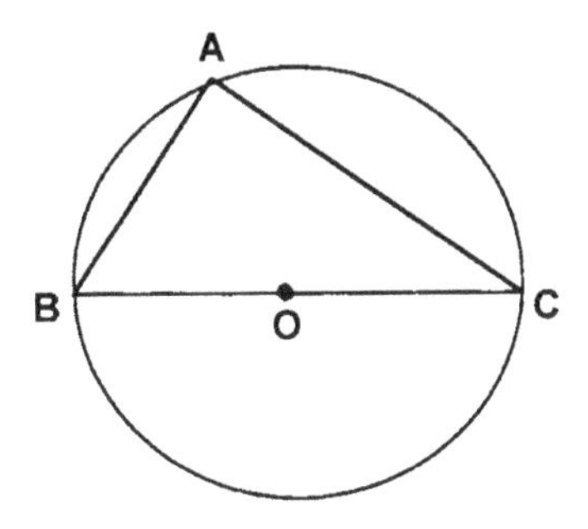

You may want to use these facts to develop your proof:

- The sum of the three angles in a triangle is 180 degrees.
- The diameter is a straight line made up of two radii (**OC** and **OB**).
- The radii of a circle are all equal.
- An isosceles triangle is a triangle with two equal sides.
- The angles in an isosceles triangle opposite the equal sides are equal.

Inductive Reasoning

9. Multiply several numbers by 9. Add up the digits of each product. If the result of the addition is a number that is more than one digit, add up the digits. Keep doing this until you get a one-digit number. For example:

$9 \times 50 = 450$

Add the digits of the product: $4 + 5 + 0 = 9$

$9 \times 43 = 387$

Add the digits of the product: $3 + 8 + 7 = 18$ (two digits)

Add the digits of the sum: $1 + 8 = 9$

$9 \times 693 = 6{,}237$

Add the digits of the product: $6 + 2 + 3 + 7 = 18$ (two digits)

Add the digits of the sum: $1 + 8 = 9$

Do you detect an emerging pattern here? If so, what is it?

10. Now, use the pattern discovered in the previous problem to determine which of the following numbers is a multiple of 9:

A. 477

B. 648

C. 8,765

D. 738

E. 9,878

11. Consider the squares of the numbers from 1 to 20:

$1^2 = 1 \times 1 = 1$

$2^2 = 2 \times 2 = 4$

$3^2 = 3 \times 3 = 9$

$4^2 = 4 \times 4 = 16$

$5^2 = 5 \times 5 = 25$

...

$20^2 = 20 \times 20 = 400$

Do you detect a pattern here? If so, what can you predict about the square of 22 and the square of 23?

Insight Thinking

12. Recall the previous Nine-Dot Puzzle. It was solved with four lines. Can it be solved with only three straight lines? That is, can you connect the nine dots without lifting your pencil, using only three straight lines?

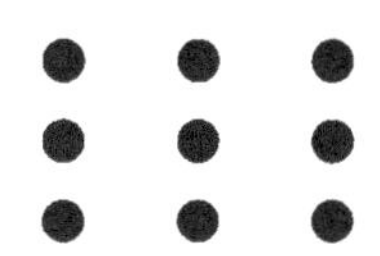

13. In the following version of the puzzle, there are twelve dots. Connect them without lifting your pencil. What is the least number of straight lines required to do so?

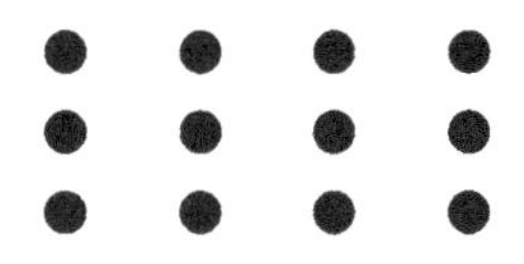

14. Finally, connect sixteen dots without lifting your pencil. What is the least number of straight lines required this time around?

Further Reading

The following list contains collections of puzzles and complementary treatments of the role of puzzles in the development of mathematics.

Averbach, Bonnie, and Orin Chein. *Problem Solving through Recreational Mathematics.* New York: Dover, 1980.

Ball, W. W. Rouse. *Mathematical Recreations and Essays,* 12th edition, revised by H. S. M. Coxeter. Toronto: University of Toronto Press, 1972.

Casti, John L. *Mathematical Mountaintops: The Five Most Famous Problems of All Time.* Oxford, N.Y.: Oxford University Press, 2001.

Costello, Matthew J. *The Greatest Puzzles of All Time.* New York: Dover, 1988.

Danesi, Marcel. *The Puzzle Instinct: The Meaning of Puzzles in Human Life.* Bloomington: Indiana University Press, 2002.

De Morgan, Augustus. *A Budget of Paradoxes.* New York: Dover, 1954.

Devlin, Keith. *The Millennium Problems: The Seven Greatest Unsolved Mathematical Puzzles of Our Time.* New York: Basic Books, 2002.

Dörrie, Heinrich. *100 Great Problems of Elementary Mathematics.* New York: Dover, 1965.

Dudeney, Henry E. *The Canterbury Puzzles and Other Curious Problems.* New York: Dover, 1958.

———. *538 Puzzles and Curious Problems.* New York: Scribner, 1967.

———. *Modern Puzzles and How to Solve Them.* London: Nelson, 1919.

Eiss, H. E. *Dictionary of Mathematical Games, Puzzles, and Amusements.* New York: Greenwood, 1988.

Falletta, Nicholas. *The Paradoxicon: A Collection of Contradictory Challenges, Problematical Puzzles, and Impossible Illustrations.* New York: John Wiley & Sons, 1990.

Gardner, Martin. *Aha! Insight!* New York: Scientific American, 1979.

———. *The Colossal Book of Mathematics.* New York: Norton, 2001.

———. *Gotcha! Paradoxes to Puzzle and Delight.* San Francisco: Freeman, 1982.

———. *The Last Recreations: Hydras, Eggs, and Other Mathematical Mystifications.* New York: Copernicus, 1997.

———. *Mathematics, Magic, and Mystery.* New York: Dover, 1956.

———. *Riddles of the Sphinx and Other Mathematical Tales.* Washington, D.C.: Mathematical Association of America, 1987.

Hovanec, Helene. *The Puzzlers' Paradise: From the Garden of Eden to the Computer Age.* New York: Paddington Press, 1978.

Kasner, Edward, and John Newman. *Mathematics and the Imagination.* New York: Simon and Schuster, 1940.

Moscovich, Ivan. *Puzzles, Paradoxes, Illusions and Games.* New York: Workman, 2001.

Olivastro, Dominic. *Ancient Puzzles: Classic Brainteasers and Other Timeless Mathematical Games of the Last 10 Centuries.* New York: Bantam, 1993.

Taylor, A. *English Riddles from Oral Tradition.* Berkeley, Calif.: University of California Press, 1951.

Townsend, Charles B. *The World's Best Puzzles.* New York: Sterling, 1986.

Van Delft, P., and J. Botermans. *Creative Puzzles of the World.* Berkeley, Calif.: Key Curriculum Press, 1995.

Wells, David. *The Penguin Book of Curious and Interesting Puzzles.* Harmondsworth, U.K.: Penguin, 1992.

Zebrowski, E. *A History of the Circle: Mathematical Reasoning and the Physical Universe.* New Brunswick, N.J.: Rutgers University Press, 1999.

2

Alcuin's River-Crossing Puzzle

In all chaos there is a cosmos,
in all disorder a secret order.

CARL JUNG (1875–1961)

PUZZLES ARE ADDICTIVE. Just ask anyone who does crossword puzzles on a daily basis or belongs to a chess or Scrabble club. Cases of puzzle addiction, in fact, fill the annals of clinical psychology. In 1925, a Broadway play called *Puzzles of 1925* satirized puzzle addiction in a hilarious way. The heart of the play featured a scene in a "Crossword Sanitarium," where people driven insane by their obsession over crossword puzzles were confined.

One of the first puzzle addicts in history was none other than Charlemagne (742–814), the founder of the Holy Roman Empire, who became so obsessed with puzzles that he hired an expert puzzle maker to create them specifically for him. The person he selected for the job was the famous English scholar and ecclesiastic Alcuin. The resourceful Alcuin put fifty-six of the puzzles he invented for Charlemagne into an instructional manual, titled *Propositiones ad acuendos juvenes* ("Problems to Sharpen the Young"), in an attempt to get medieval youths interested in mathematics.

One puzzle in that anthology, known as the River-Crossing Puzzle, qualifies as among the ten greatest of all time. Not only is it included in virtually all the classic puzzle anthologies, but many mathematical historians consider the idea pattern on which it is constructed to be the key insight that led, centuries later, to the establishment of a branch of mathematics known as **combinatorics**, which deals essentially with the structure of arrangements. It attempts to determine how things can be grouped,

ALCUIN (735–804)

Alcuin was a renowned medieval scholar, teacher, and writer. He studied at the cloister-school of York, the center of learning in England during his era. Alcuin became an adviser to Emperor Charlemagne in 782. In 796, Charlemagne made him the abbot of St. Martin at Tours, in France. In that post, Alcuin helped to spread the achievements of Anglo-Saxon scholarship throughout Europe, bringing about the revival of learning known as the Carolingian Renaissance.

Alcuin's puzzle anthology became widely known in the medieval world, and many of its puzzles continue to find their way, in one version or other, into contemporary collections. All of them require a high degree of ingenuity to solve.

counted, or organized in some systematic way. Although the same idea pattern is found in the puzzle traditions of different cultures, Alcuin's version became widely known to mathematicians. Remarkably, Alcuin's simple puzzle has had important implications for the study of logic and for the design and the operation of computers.

The Puzzle

A common rendition of the River-Crossing Puzzle is the following one:

> A traveler comes to a riverbank with a wolf, a goat, and a very large head of cabbage. To his chagrin, he notes that there is only one boat for crossing over, which can carry no more than two—the traveler and one of the two animals or the cabbage. As the traveler knows, if left alone together, the goat will eat the cabbage and the wolf will eat the goat. The wolf does not eat cabbage. How does the traveler transport his animals and his cabbage to the other side intact, in a minimum number of back-and-forth trips?

The traveler starts by taking the goat to the other side, leaving the wolf with the cabbage on the original side. He rows back alone. He then takes the wolf across, leaving the cabbage by itself on the original side. On the other side he leaves the wolf and rows back with the goat. On the original side, he then leaves the goat and takes the cabbage across. He rows back alone, leaving the wolf and the cabbage safely together on the other side. He picks up the goat on the original side and rows across. When he gets to the other side, he has his wolf, goat, and cabbage intact and so can continue on his journey. The whole process took seven back-and-forth crossings.

Here is a step-by-step modeling of the solution. The "initial-state" on both sides of the river, before the traveler starts rowing back and forth, can be shown as follows (W = wolf, G = goat, C = cabbage, T = traveler). This state can be represented as a "0" step, because no rowing is involved:

	On the Original Side	**On the Boat**	**On the Other Side**
0.	W G C T	_ _	_ _ _ _

The traveler begins by transporting the goat over on the boat, leaving the wolf and the cabbage alone on the original side without any problems. This constitutes the first step in the solution:

	On the Original Side	**On the Boat**	**On the Other Side**
0.	W G C T	_ _	_ _ _ _
1.	W _ C _	T G →	_ _ _ _

The traveler deposits the goat on the other side and then goes back alone. This completes the first round trip, adding a second step to the solution:

	On the Original Side	**On the Boat**	**On the Other Side**
0.	W G C T	_ _	_ _ _ _
1.	W _ C _	T G →	_ _ _ _
2.	W _ C _	← T _	_ G _ _

From the original side, the traveler picks up the wolf and rows across, leaving the cabbage by itself. This is the third step:

	On the Original Side	**On the Boat**	**On the Other Side**
0.	W G C T	_ _	_ _ _ _
1.	W _ C _	T G →	_ _ _ _
2.	W _ C _	← T _	_ G _ _
3.	_ _ C _	T W →	_ G _ _

Once on the other side, the traveler cannot leave the wolf and the goat alone, for the former would eat the latter. So, he brings the goat along for the ride, leaving the wolf by itself. This constitutes the fourth step:

	On the Original Side	**On the Boat**	**On the Other Side**
0.	W G C T	_ _	_ _ _ _
1.	W _ C _	T G →	_ _ _ _
2.	W _ C _	← T _	_ G _ _
3.	_ _ C _	T W →	_ G _ _
4.	_ _ C _	← T G	W _ _ _

Back on the original side, the traveler leaves the goat and takes the cabbage with him over to the wolf. In this way, he avoids leaving the goat and the cabbage together. This is the fifth step:

	On the Original Side	**On the Boat**	**On the Other Side**
0.	W G C T	_ _	_ _ _ _
1.	W _ C _	T G →	_ _ _ _
2.	W _ C _	← T _	_ G _ _
3.	_ _ C _	T W →	_ G _ _
4.	_ _ C _	← T G	W _ _ _
5.	_ G _ _	T C →	W _ _ _

The traveler then deposits the cabbage on the other side. He goes back alone, leaving the wolf and the cabbage safely together. This is the sixth step:

	On the Original Side	**On the Boat**	**On the Other Side**
0.	W G C T	_ _	_ _ _ _
1.	W _ C _	T G →	_ _ _ _
2.	W _ C _	← T _	_ G _ _
3.	_ _ C _	T W →	_ G _ _
4.	_ _ C _	← T G	W _ _ _
5.	_ G _ _	T C →	W _ _ _
6.	_ G _ _	← T _	W _ C _

For his last trip across to the other side, the traveler brings the goat with him on the boat. This is the seventh step in the model:

	On the Original Side	On the Boat	On the Other Side
0.	W G C T	_ _	_ _ _ _
1.	W _ C _	T G →	_ _ _ _
2.	W _ C _	← T _	_ G _ _
3.	_ _ C _	T W →	_ G _ _
4.	_ _ C _	← T G	W _ _ _
5.	_ G _ _	T C →	W _ _ _
6.	_ G _ _	← T _	W _ C _
7.	_ _ _ _	T G →	W _ C _

When the traveler reaches the other side, he can continue happily on his journey with his wolf, goat, and cabbage. This is the "end-state" in the model. Like the initial-state, "0" can be used to represent it, since no rowing is involved. The complete solution model is given as follows:

	On the Original Side	On the Boat	On the Other Side
0.	W G C T	_ _	_ _ _ _
1.	W _ C _	T G →	_ _ _ _
2.	W _ C _	← T _	_ G _ _
3.	_ _ C _	T W →	_ G _ _
4.	_ _ C _	← T G	W _ _ _
5.	_ G _ _	T C →	W _ _ _
6.	_ G _ _	← T _	W _ C _
7.	_ _ _ _	T G →	W _ C _
0.	_ _ _ _	_ _	W G C T

A slightly different seven-step solution is also possible. In this case, too, the traveler starts by bringing over the goat first. The difference between this and the previous solution occurs in the third, fourth, and fifth steps:

	On the Original Side	On the Boat	On the Other Side
0.	W G C T	_ _	_ _ _ _
1.	W _ C _	T G →	_ _ _ _

2.	W _ C _	← T _	_ G _ _
3.	W _ _ _	T C →	_ G _ _
4.	W _ _ _	← T G	_ _ C _
5.	_ G _ _	T W →	_ _ C _
6.	_ G _ _	← T _	W _ C _
7.	_ _ _ _	T G →	W _ C _
0.	_ _ _ _	_ _	W G C T

Readers who may prefer to actually carry out the previous crossings in a concrete way can do so with, say, a box to represent the boat and four slips of paper representing the wolf, the goat, the cabbage, and the traveler (W, G, C, T).

An interesting version of the puzzle was devised by the Italian mathematician Niccolò Tartaglia in the sixteenth century. It featured three brides and their jealous husbands:

> Three beautiful brides with their husbands come to a river. The small boat that will take them across holds only two people. To avoid any compromising situations, the crossings are to be so arranged that no woman shall be left alone with a man unless her husband is present. How can this be done, if any man or woman can be the rower?

Nine crossings are required. As we did earlier, we can easily model the solution by letting H stand for a husband and W for a wife, using subscript numbers to indicate who is married to whom—H_1 and W_1 will thus represent one husband and wife pair, H_2 and W_2 a second pair, and H_3 and W_3 a

NICCOLÒ FONTANA TARTAGLIA (c. 1499–1557)

Tartaglia was born in Venice, where he was widely known as a scientist and a mathematician. His most notable work was the *Nova Scientia*, in which he discussed the motion of heavenly bodies and the trajectory of projectiles.

Tartaglia was also the first one to devise an algorithm (a step-by-step procedure) for solving cubic equations in 1541. These are equations in which one of the variables is raised to the power of three: for example, $x^3 + 29x^2 = 145$. But it was his rival Girolamo Cardano (1501–1576) who became famous for the solution, which, as some historians claim, he probably stole from Tartaglia.

third pair. The basic idea is to avoid having a W and an H with different subscript numbers alone together (on the boat or on a side). Thus, for example, a pairing such as H_1 and W_2 on the boat is inappropriate, since it would constitute a pair in which a woman (W_2) is alone with a man (H_1) who is not her husband—which the puzzle forbids. All other pairings are allowable. One possible nine-step model is given as follows, without commentary. There are others. Readers may again wish to use a box for the boat and six pieces of paper for the people, labeling them H_1, H_2, H_3, W_1, W_2, and W_3, and then physically carry out each step as shown in the model:

	On the Original Side	**On the Boat**	**On the Other Side**
0.	H_1 W_1 H_2 W_2 H_3 W_3		
1.	H_2 W_2 H_3 W_3	H_1 W_1 →	
2.	H_2 W_2 H_3 W_3	← W_1	H_1
3.	H_2 H_3 W_3	W_1 W_2 →	H_1
4.	H_2 H_3 W_3	← W_2	H_1 W_1
5.	H_3 W_3	H_2 W_2 →	H_1 W_1
6.	H_3 W_3	← W_2	H_1 W_1 H_2
7.	H_3	W_2 W_3 →	H_1 W_1 H_2
8.	H_3	← W_3	H_1 W_1 H_2 W_2
9.		H_3 W_3 →	H_1 W_1 H_2 W_2
0.			H_1 W_1 H_2 W_2 H_3 W_3

More complicated versions of the River-Crossing Puzzle, involving more people and animals, can easily be constructed. However, not all are solvable. For instance, as the well-known puzzlists Sam Loyd (1841–1911) and Henry E. Dudeney (1847–1930) discovered, it is impossible to arrive at a solution under the conditions stipulated by Tartaglia's puzzle for four couples. In such a case, a solution is possible only if there is an island in midstream for use as a "transit stop." The Loyd-Dudeney puzzle is assigned as an exploration exercise, further on.

Mathematical Annotations

A problem based on a certain arrangement of things—animals, married couples, letters of the alphabet—can be studied systematically and modeled accurately. That is the main lesson to be learned from Alcuin's puzzle. Mathematical modeling is the activity of representing all kinds of patterns—numerical, geometrical, combinatory, and so on.

The Josephus and Kirkman School Girl Puzzles

Puzzles in arrangement abound. All require a large dose of insight thinking to solve. Consider two other famous examples. The first one is called the Josephus Puzzle, after the Jewish historian Josephus of the first century A.D., who supposedly saved his own life by coming up with the correct solution. Here is a version of that puzzle:

> There are fifteen tyrants (T) and fifteen helpless citizens (C) on a ship—way too many for the size of the ship. So, it is decided that the tyrants must be thrown overboard to prevent the ship from sinking. A mythical beast, who cannot distinguish between tyrants and citizens, has been let loose on the ship to throw people overboard. The beast has been trained to throw over every ninth person seated in a circle. How can the people on board be arranged in a circle so that the beast can do the job it is expected to do?

The beast starts at the "C" at the top of the circle shown. The ninth person from the start is a "T." So he is thrown overboard. The ninth person after that is also a "T." He, too, is thrown overboard. And so on. The circular seating arrangement shown as follows thus guarantees that every tyrant is thrown overboard, while all the citizens are saved, as readers can confirm for themselves.

Versions of the Josephus Puzzle are found in different cultures throughout the world. The puzzle was studied by famous mathematicians, including Leonhard Euler (whom we will meet in chapter 4), because it constitutes, in puzzle form, a miniature model for investigating more complex problems in systematic arrangement—an area of study that now goes under the rubric of **systems analysis**.

The second puzzle is called Kirkman's School Girl Puzzle, after the notable amateur mathematician Thomas Penyngton Kirkman, who posed it in 1847. It, too, has had important implications for the study of arrangement and, especially, for matrix theory. A **matrix** in mathematics is an arrangement of numbers (or symbols) distributed in columns and rows according to some pattern:

> How can fifteen girls walk in five rows of three each for seven days so that no girl walks with any other girl in the same triplet more than once?

The puzzle can be reformulated more prosaically as follows: How can the first fifteen digits, from 0 to 14 (each one representing a specific girl), be divided into seven matrices (each matrix corresponding to a day of the week), consisting of five sets of triplets, so that no two digits appear in the same row more than once in any of the seven matrices? This puzzle is assigned as an exercise in the Explorations section, further on.

In sum, the River-Crossing, Josephus, and School Girl puzzles were all important for laying down the conceptual foundations on which the science of combinatorics was built in the nineteenth century. Combinatorics studies the mathematical structure of arrangements.

Combinatorics

Combinatorics aims essentially to answer the following question: what ordered arrangements of a set of objects are possible under certain conditions? To grasp what this entails, let's return to Alcuin's puzzle, slightly changing the conditions in it. This time, the traveler finds a boat that has four seats in total—one seat for him and three other seats in a row, in which he can put the wolf, the goat, and the cabbage. One trip across will suffice this time. The question is:

> In how many possible arrangements can the traveler seat the wolf, the goat, and the cabbage in his boat?

Let's label the three seats 1, 2, and 3. Consider the first seat. The traveler could put any one of the three (W = wolf, G = goat, C = cabbage) in it:

Seat 1	Seat 2	Seat 3
↓	↓	↓
W	__	__
G	__	__
C	__	__

For each of the previous placements in seat 1, the traveler can put either one of the remaining two in seat 2. For example, with W in seat 1, he can put G or C in seat 2; with G in seat 1, he can put W or C in seat 2; and with C in seat 1, he can put W or G in seat 2. The possible choices the traveler has for putting the wolf, the goat, and the cabbage in the first two seats are summarized as follows:

Seat 1		Seat 2		Outcome	Seat 3
↓		↓		↓	↓
W	with	G	→	W G	__
W	with	C	→	W C	__
G	with	W	→	G W	__
G	with	C	→	G C	__
C	with	W	→	C W	__
C	with	G	→	C G	__

Note that the total number of potential pairings so far (shown in the previous "Outcome" column)—WG, WC, GW, GC, CW, CG—is $3 \times 2 = 6$. This expresses in arithmetical form the fact that for each of the three things the traveler puts in seat 1, he can place two others in seat 2, producing a total of six pairs.

Now, for each of these six pairs, the traveler has one choice left for putting something in seat 3. For example, with the pair WG in seats 1 and 2 respectively, the traveler has C left for seat 3; with WC in seats 1 and 2 respectively, he has G left for seat 3; and so on. Here is a summary of all the possible arrangements of the wolf, the goat, and the cabbage on the boat:

Seat 1		Seat 2		Outcome		Seat 3		Final Outcome
↓		↓		↓		↓		↓
W	with	G	→	W G	with	C	→	W G C
W	with	C	→	W C	with	G	→	W C G
G	with	W	→	G W	with	C	→	G W C
G	with	C	→	G C	with	W	→	G C W
C	with	W	→	C W	with	G	→	C W G
C	with	G	→	C G	with	W	→	C G W

Note again that the total number of potential pairings (shown in the previous "Final Outcome" column)—WGC, WCG, GWC, GCW, CWG, and CGW—is $3 \times 2 \times 1 = 6$. As before, this expresses in arithmetical form the fact that for each of the three things the traveler puts in seat 1, he can place two in seat 2, and one in seat 3, producing a total of six triplet arrangements. The arrangements are called permutations, as readers may recall from their school mathematics. A **permutation** is a grouping of elements with regard to their order. For example, the result of permuting two letters, A and B, is AB and BA. The pairs consist of the same two elements, but the order is different.

Now, let's include the traveler (T) himself in the arrangement planning:

> In how many possible arrangements can the traveler seat himself, the wolf, the goat, and the cabbage in the four seats on the boat?

In this case, there are $4 \times 3 \times 2 \times 1 = 24$ possible arrangements. The first digit 4 refers to the fact that any one of the four—the wolf (W), the goat (G), the cabbage (C), or the traveler (T)—can occupy seat 1. The second digit 3 tells us that for each of the four possibilities for seat 1, there are three ways in which seat 2 can be occupied. This produces $4 \times 3 = 12$ permutations for the first two seats. The third digit tells us that for each of the twelve permutations, there are two ways in which seat 3 can be occupied, for a total of $4 \times 3 \times 2 = 24$ permutations. Finally, the fourth digit indicates that for each of the twenty-four permutations so far, there is only one way in which seat 4 can be occupied, for a total of $4 \times 3 \times 2 \times 1 = 24$ permutations. These permutations are given as follows (organized, for the sake of clarity, according to the occupant of the first seat):

With W in Seat 1:

1. WGCT
2. WGTC
3. WCGT
4. WCTG
5. WTGC
6. WTCG

With G in Seat 1:

7. GWCT
8. GWTC
9. GCWT
10. GCTW
11. GTWC
12. GTCW

With C in Seat 1:

13. CWGT
14. CWTG
15. CGWT
16. CGTW
17. CTWG
18. CTGW

With T in Seat 1:

19. TWGC
20. TWCG
21. TGWC
22. TGCW
23. TCWG
24. TCGW

If there were five places to be occupied on the boat and five things to fill them, there would be $5 \times 4 \times 3 \times 2 \times 1$ possible arrangements, or permutations, of the things (as readers can confirm for themselves); if there were six places and six things, there would be $6 \times 5 \times 4 \times 3 \times 2 \times 1$ permutations; and so on. Do you see the pattern? If there were n places to be occupied and n

objects to fill them, there would be $n \times (n-1) \times (n-2) \times \ldots 1$ permutations. This is known as a **factorial**, and is represented by the symbol $n!$:

$$n! = n \times (n-1) \times (n-2) \times \ldots \times 1.$$

This formula generalizes the fact that the first position can be filled with n objects, the second position with one less than the total number of objects $(n-1)$, the third with two less than the total number $(n-2)$, and so on, down to a single possibility for the last position. Here are some examples of factorials:

$$4! = 4 \times 3 \times 2 \times 1$$
$$5! = 5 \times 4 \times 3 \times 2 \times 1$$
$$9! = 9 \times 8 \times 7 \times 6 \times 5 \times 4 \times 3 \times 2 \times 1$$
$$\ldots$$
$$n! = n \times (n-1) \times (n-2) \times \ldots \times 1$$

Let's explore the concept of permutation a little further. Suppose that some of the objects to be arranged are the same. Here's an example of this kind of problem:

> How many five-digit numerals can you construct using the digits 1, 1, 2, 3, and 4?

Here we have five objects (digits), two of which are indistinguishable—the two 1s. This means that some arrangements will turn out to be exactly the same. These must therefore be "filtered out," so to speak. To do this, let's assign a subscript to the two 1s, in order to keep track of them. The five digits, therefore, can be rewritten as follows:

$$1_1, 1_2, 2, 3, \text{ and } 4.$$

A total number of 120 five-digit numerals can be made:

$$5! = 5 \times 4 \times 3 \times 2 \times 1 = 120.$$

However, some of these will be indistinguishable when the subscripts are removed, as examples 1 and 2 show:

1. $1_1 2 1_2 34 = 12{,}134$
2. $1_2 2 1_1 34 = 12{,}134$

Now, how many of these cases exist among the 120 numerals? There will be 2! of them. Why? Because that is how many permutations there are of the two digits 1_1 and 1_2 when considered in isolation from their occurrence with the other digits in the numerals. Patient readers may wish to verify this for themselves by constructing the 120 permutations with the subscripted numbers and then canceling out those that are indistinguishable. The canceling out of the indistinguishable cases is, of course, a case of division, whereby the 2! is divided into 5!:

$$\frac{5!}{2!} = \frac{5 \times 4 \times 3 \times \cancel{(2 \times 1)}}{\cancel{(2 \times 1)}} = 5 \times 4 \times 3 = 60.$$

We can thus make sixty distinct five-digit numerals. In general, the number of distinct permutations of n objects, among which there are r indistinguishable cases, is

$$\frac{n!}{r!} \quad (r = \text{number of indistinguishable cases}).$$

Now, let's consider a situation in which we have more people than positions in which to place them:

> Suppose you had to choose one person to be president, one vice president, and one secretary from a committee of ten people. In how many ways could you fill those three positions?

Of course, ten people can be selected for the position of president; for each of these, any one of the remaining nine can be selected for the vice presidency; and for each of the previously selected pairs, we can choose from among eight people to fill the remaining position of secretary. So, the number of total permutations in this case is $10 \times 9 \times 8 = 720$.

Can this solution be generalized? Notice that the answer to $10 \times 9 \times 8$ is, in effect, 10! with the last seven factors canceled from it:

$$10 \times 9 \times 8 \cancel{(\times 7 \times 6 \times 5 \times 4 \times 3 \times 2 \times 1)}.$$

This result is produced by dividing 7! ($= 7 \times 6 \times 5 \times 4 \times 3 \times 2 \times 1$) into 10!—which is shown as follows:

$$\frac{10!}{7!} = \frac{10 \times 9 \times 8 \times \cancel{(7 \times 6 \times 5 \times 4 \times 3 \times 2 \times 1)}}{\cancel{(7 \times 6 \times 5 \times 4 \times 3 \times 2 \times 1)}} = 10 \times 9 \times 8 = 720.$$

Note that the denominator 7! in the previous fraction can cleverly be rewritten as $(10 - 3)!$ and that the 3 in parentheses stands for the number of positions to be filled. This resourceful bit of rewriting provides the relevant insight. In general, if the numerator is $n!$ then the denominator will be $(n - r)!$ where r stands for the number of positions to be filled:

$$\frac{n!}{(n-r)!}.$$

This formula now allows us to solve any problem that requires us to put n objects in r positions easily. Another way to phrase it is that the formula allows us to permute n objects taken r at a time.

Before leaving the field of elementary combinatorics, let's look at one final type of arrangement pattern. Suppose we wanted to select a three-member subcommittee from the ten candidates. Are there also 720 ways to do this? The answer is no, because order does not matter in this case. For example, let's say that three of the people to be chosen are named Chris, Lucy, and Rachel. There are 3! ways ($3! = 3 \times 2 \times 1 = 6$) to select these three specific people to fill the three positions on the subcommittee:

President	**Vice President**	**Secretary**
↓	↓	↓
Chris	Lucy	Rachel
Chris	Rachel	Lucy
Lucy	Chris	Rachel
Lucy	Rachel	Chris
Rachel	Chris	Lucy
Rachel	Lucy	Chris

When making a subcommittee, however, the order of the selections is irrelevant. It only matters that three are chosen. It is irrelevant, for instance, whether the order is (1) Chris, Lucy, Rachel or (6) Rachel, Lucy, Chris. The previous six permutations, therefore, can be reduced to a combination of three distinct people. That is why this type of arrangement is called a combination, rather than a permutation. A **combination** can be defined as an arrangement that is put together with no regard to order. In the case of a three-member subcommittee, the redundant permutations—3!—must be eliminated from the 720 possible selections that can be made. How many of these are there? To determine this, we simply cancel out the redundant permutations from 720 through division:

$$\frac{720}{3!} = \frac{720}{6} = 120.$$

A SUMMARY OF FORMULAS

The permutation of n objects: $n! = n \times (n-1) \times (n-2) \times \ldots \times 1$

The permutation of n objects, in which r cases are indistinguishable: $n!/r!$

The permutation of n objects, taken r at a time: $n!/(n-r)!$

The combination of n objects, taken r at a time: $n!/(n-r)!\,r!$

Note that the denominator in this case is again $r!$—the number of positions to be filled. Note as well that the 720 in the numerator was produced by our previous formula—namely, $n!/(n-r)!$. So, the general formula for a combination is the previous formula divided by $r!$:

$$\frac{n!}{(n-r)!\,r!}.$$

Reflections

The River-Crossing, Josephus, and School Girl puzzles are probes of combinatory pattern. In simple yet elegant ways, they exemplify what mathematical inquiry is all about. As the great German mathematician Gottfried Leibniz (1646–1716) so aptly put it, mathematics is an *ars combinatoria*, a "combinatory art."

PYTHAGORAS (c. 582–500 B.C.)

Pythagoras is best known for discovering the theorem named after him. Around 529 B.C., he settled in Crotona (Italy). There, he founded a secret society among the aristocrats. People in the inner circle were called mathematikoi, which meant "those trained in science." Suspicious of the society, citizens killed most of its members in a political uprising. Historians do not know whether Pythagoras left the city before the outbreak of violence or whether he was killed in it. The Pythagorean society continued on for a while after the slaughter, disappearing from history in the 400s B.C.

The Greek philosopher Pythagoras founded mathematics, in fact, as the science of pattern. His greatest discovery was the theorem that bears his name (the **Pythagorean theorem**). The theorem states that the square of the **hypotenuse** (the side of a triangle opposite the right angle) of a **right-angled triangle** (a triangle containing a 90-degree angle) is equal to the sum of the squares of the other two sides. The hypotenuse is the side c opposite the right angle in the figure below. More will be said about this theorem in chapter 5.

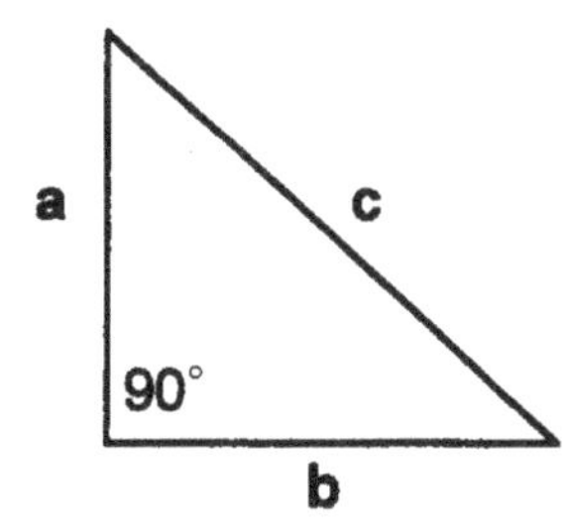

The Pythagoreans firmly believed that mathematical theorems (such as the one they themselves proved) held the secrets of the universe. The cosmos speaks to us in a numerical language, they claimed. For this reason, the objective of mathematics must surely be one of the most important of all human objectives—to decode the grammar of that language.

Explorations

Crossings, Arrangements, Pairings

15. Let's start with a simple version of Tartaglia's puzzle, to get the mental engines running, so to speak, adding a slightly different twist to it. How many back-and-forth crossings are required for only two couples if (1) the small boat that is to take them across holds only two people, (2) to avoid any compromising situations, the crossings are to be so organized that no woman shall be left with a man unless her husband is present, and (3) the two women can never be left alone together (on either side or on the boat)?

16. Now, determine the number of complete back-and-forth crossings needed for four husband-and-wife pairs if, again, the small boat that is to take them across holds only two people, and the crossings are to be so organized that no woman shall be left with a man unless her husband is present. Note that the solution is possible only if there is an island in midstream for people to use as a "transit stop." Trips that involve going to the island and doubling back from it do not count as "complete" crossings. In this version, two or more women can be left alone together anywhere and at any time.

17. Solve Kirkman's puzzle: how can fifteen girls walk in five rows of three each for seven days so that no girl walks with any other girl in the same row (triplet) more than once?

18. There are twenty billiard balls, ten white and ten black, in a box. After you put a blindfold on, what is the least number of draws you must make to be sure of having a pair of balls of matching color—that is, two white balls or two black balls?

19. Now, what is the least number of draws you must make (always with a blindfold on) to be sure of having a pair of balls of matching color, when the box contains

A. ten white, ten black, and ten green balls

B. ten white, ten black, ten green, and ten yellow balls

C. ten white, ten black, ten green, ten yellow, and ten red balls?

Do you detect a pattern? If so, what is it?

20. Would the same pattern apply if the number of balls varied: for example, ten white, eight black, and four green?

21. If there are six pairs of black gloves and six pairs of white gloves in a box, all mixed up, what is the least number of draws you must make, with a blindfold on, to be sure of having a matching pair of black or white gloves?

22. Perhaps the most ingenious of all the puzzles in this genre is the one devised by Lewis Carroll (1832–1898), the great puzzlist and the author of the classic children's books *Alice's Adventures in Wonderland* (1865) and *Through the Looking-Glass* (1872). A bag contains one counter, which is either white or black. A white counter is put in, the bag is shaken, and a counter drawn out, which proves to be white. What are the chances of drawing a white counter?

Combinatorics

23. If there are three different routes from Sarah's to Bill's house and four different routes from Bill's to Shirley's house, how many routes are there from Sarah's to Shirley's house that go through Bill's house?

24. A club has twenty members. It is electing a president and a vice president. How many different outcomes of the election are possible? What if only two candidates, call them Brenda and Heather, are allowed to be elected president?

25. Alex wants to make soup with exactly five different vegetables. If he has twelve vegetables from which to choose, how many different soups can he make?

Further Reading

Ascher, M. "A River-Crossing Problem in Cross-Cultural Perspective." *Mathematics Magazine* 63 (1990): 26–29.

Biggs, N. L. "The Roots of Combinatorics." *Historia Mathematica* 6 (1979): 109–36.

Gerdes, P. "On Mathematics in the History of Sub-Saharan Africa." *Historia Mathematica* 21 (1994): 23–45.

Pressman, Ian, and David Singmaster. "The Jealous Husbands and the Missionaries and Cannibals." *Mathematical Gazette* 73 (1989): 73–81.

Primrose, E. J. F. "Kirkman's Schoolgirls in Modern Dress." *Mathematical Gazette* 60 (1976): 292–93.

Fibonacci's Rabbit Puzzle

The more we know the more fantastic the world becomes
and the profounder the surrounding darkness.

ALDOUS HUXLEY (1894–1963)

TODAY WE OFTEN REFER to the medieval European era as the "Dark Ages," even though, as it turns out, it was a much more "enlightened" period of science than we commonly think. Indeed, significant discoveries in physics, chemistry, and astronomy were made by medieval scholars, despite considerable resistance to their efforts by the ruling religious oligarchy of the era.

But in one area of medieval knowledge, the expression "Dark Ages" may be an appropriate one, after all. Until the early thirteenth century, little progress had been made in mathematics—not because of any opposition from religious authorities, and certainly not for any lack of ingenuity, but because such progress was likely hampered by the cumbersome and inefficient numeration system in use at the time—the **Roman numeral** one—which was based on seven alphabet letters, having specific numerical values:

I = one

V = five

X = ten

L = fifty

C = one hundred

D = five hundred

M = one thousand

To grasp how unwieldy that system was, consider how the numeral "two thousand two hundred fifty-three" was put together:

MMCCLIII = two thousand two hundred fifty-three.

Now, compare the Roman numeral with the one we use today:

2,253 = two thousand two hundred fifty-three.

Ours is clearly much easier to read, because the principle that is used to construct it is simple—the position of each digit in the numeral indicates its value as a power of ten. This is why our system is called "decimal" (from Latin *decem*, "ten"). Here is how the **decimal numeral** 2,253 is read—note that "one thousand" can be represented by 10^3 (because $10^3 = 10 \times 10 \times 10 = 1{,}000$), "one hundred" by 10^2 (because $10^2 = 10 \times 10 = 100$), "ten" by 10^1, and "one" by 10^0 (see the sidebar on exponents in chapter 1):

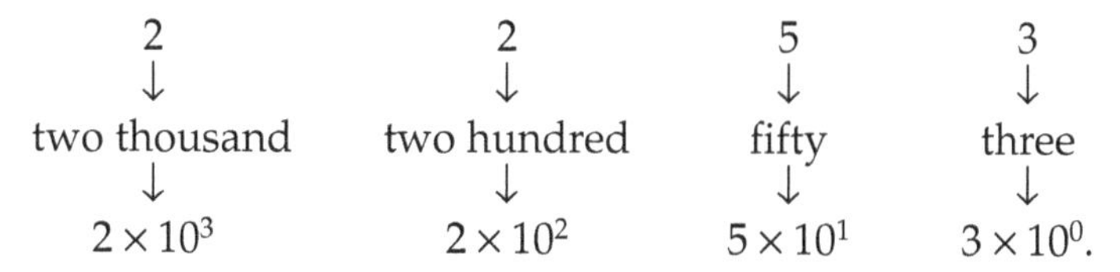

Now, imagine trying to carry out a simple arithmetical task, such as adding 2,253 + 1,337, with Roman numerals. Here's how it would look on paper:

MMCCLIII + MCCCXXXVII = MMMDXC.

The task is daunting, as readers can confirm for themselves. It is further complicated by the fact that a smaller numeral appearing before a larger one indicates that the smaller one is to be subtracted from the larger one. To wit: the numeral for "ninety" is represented by XC ("one hundred minus ten").

THE ABACUS

The **abacus** is an ancient device that was used in China and other countries to facilitate arithmetical computation. It is made with a frame containing columns of beads strung on wires or narrow wooden rods attached to the frame. The beads represent numbers.

A typical Chinese abacus has columns of beads separated by a crossbar. Each column has two beads above the crossbar and five below it. The first column on the right represents the "ones" column; the second, the "tens" column; the third, the "hundreds" column; and so on.

Clearly, it would take quite an effort to carry out the addition, keeping track of all the letter-to-number values, especially when we compare it to the minimal effort expended in performing addition with decimal numerals:

$$\begin{array}{r} 2{,}253 \\ +\ 1{,}337 \\ \hline 3{,}590\,. \end{array}$$

As mentioned, the superiority of the decimal system over the Roman one lies in the fact that it is based on the "abacus principle," whereby the position of a digit indicates its value in terms of a power of ten. The 0 digit in this system makes it possible to differentiate between numbers such as "eleven" (= 11) "one hundred and one" (= 101) and "one thousand and one" (= 1,001), without the use of additional numerals. The 0 digit in a numeral tells us, simply, that the position is "void" or "empty," since multiplying any number by 0 always yields 0:

11 = eleven

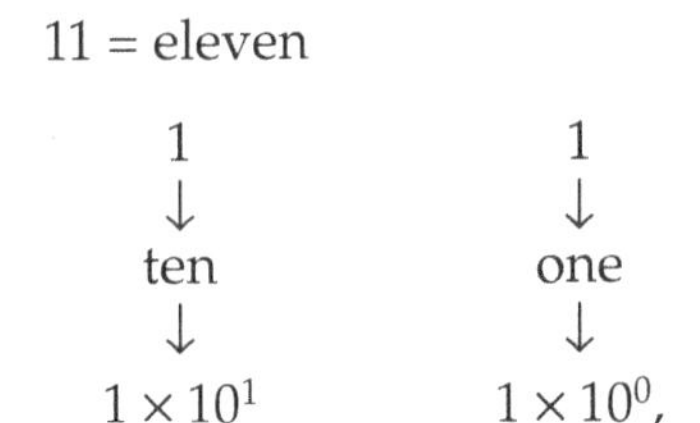

101 = one hundred and one

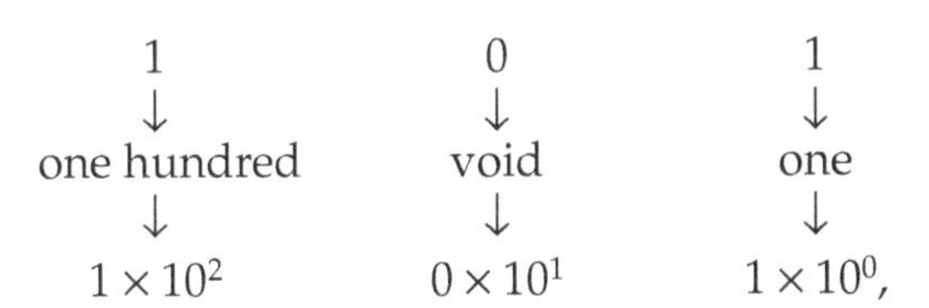

1,001 = one thousand and one

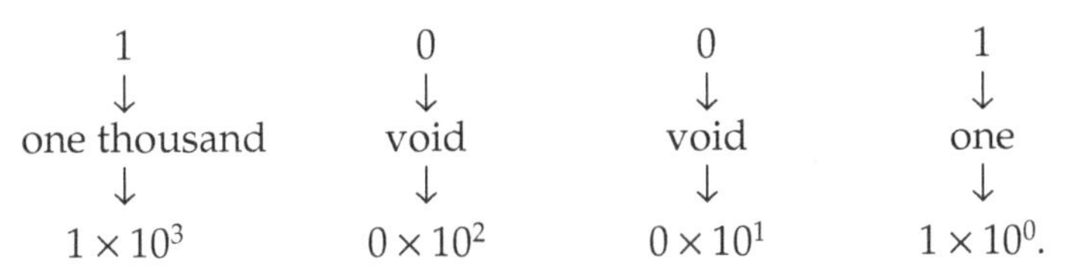

No wonder, then, that the numeration system in use in the world today is the decimal one. It was first developed by the Hindus in India in the third century B.C., and was then introduced into the Arabic world around the seventh or eighth century A.D. The Hindu-Arabic system first reached Europe in the year 1000 through the efforts of Pope Sylvester II. But it hardly got noticed at the time. It was reintroduced in a much more practical way to medieval Europeans a few centuries later by an Italian businessman named Leonardo da Pisa, Figlio di Bonacci ("Leonard from Pisa, Son of Bonacci"), known more commonly as Leonardo Fibonacci.

With the publication in 1202 of his textbook, titled appropriately the *Liber Abaci* ("The Book of the Abacus"), Fibonacci succeeded in convincing his fellow Europeans that the decimal system was far superior to the Roman one. He did this essentially by devising a series of puzzles and practical problems that could easily be solved with it. Shortly after its publication, mathematics literally "took off," becoming a thriving science throughout Europe, no doubt influencing and animating the revival of learning known as the Renaissance, which began in Italy in the early 1300s.

LEONARDO FIBONACCI
(c. 1170–1240)

Born in Pisa, Fibonacci traveled all over the Byzantine Empire. During his trips, he learned about the decimal system that was used in the Arabic world. Upon his return to Pisa in 1202, he published the *Liber Abaci* to illustrate the practicality and the efficiency of that numeral system to a European audience.

Fibonacci was so taken by Arabic culture that he wrote many things in his book from right to left, in imitation of the Arabic writing style. For example, he wrote the numerals in descending order (10, 9, 8, 7, 6, 5, 4, 3, 2, 1, 0), and he put fractions before whole numbers ($\frac{1}{2}4$ instead of $4\frac{1}{2}$).

It is in the *Liber Abaci* that the Rabbit Puzzle appears. Like Alcuin's River-Crossing Puzzle, it highlights the fact that mathematics is essentially a study of patterns. The solution to the puzzle produces a sequence of numbers that contains so many hidden patterns that to this day, people continue to flesh them out. And, if this were not enough, the "Fibonacci sequence," as it is now called, has been found to occur in nature and human affairs! Clearly, if any puzzle should make the top ten list, it is Fibonacci's Rabbit Puzzle.

The Puzzle

The puzzle is found in the third section of the *Liber Abaci*:

> A certain man put a pair of rabbits, male and female, in a very large cage. How many pairs of rabbits will be produced in that cage in a year if every month each pair produces one and only one new pair, consisting again of a male and a female, which, from the second month of its existence on, also is productive? It is assumed that none of the rabbits die in that year.

At the start, there is only one pair of rabbits in the cage—the original pair. At the end of the first month, there is still that one pair in the cage, because the puzzle states that a pair becomes productive only "from the second month of its existence on." During the second month, the pair will produce its first offspring pair. Thus, at the end of the second month, a total of two pairs—the original one and its first offspring pair—are in the cage. Let's summarize:

At the start

A pair is put into the cage.

Let's name the pair F_1, for "Fibonacci rabbit pair number 1."

After the first month

Total number of pairs in the cage: $F_1 = 1$ pair.

After the second month

F_1 has produced its first pair of offspring.

Let's name the new pair F_2.

Total number of pairs in the cage: $F_1 + F_2 = 2$ pairs.

During the third month, only the original pair, F_1 (now fully productive), gives birth to another new pair. According to the condition set down by the puzzle (a pair becomes productive from the second month of its existence), F_2 must wait a month before it, too, becomes productive. So, at the end of the third month, there are three pairs in total in the cage: the one initial pair, and the two offspring pairs that the original pair has thus far produced:

After the third month

F_1 has produced another pair of offspring.

Let's call it F_3.

F_2 has not yet produced a pair, because it has been in the cage only one month.

Total number of pairs in the cage: $F_1 + F_2 + F_3 = 3$ pairs.

Now, consider what happens during the fourth month. The original pair F_1 produces yet another pair. F_2 now produces its own first pair. F_3 has not started producing yet. Therefore, at the end of the month, a total of two newborn pairs of rabbits are added to the cage—one from F_1 and one from F_2. Altogether, at the end of the month there are the previous three pairs plus the two newborn ones, making a total of five pairs in the cage:

After the fourth month

F_1 has produced another pair of offspring.

Let's call it F_4.

F_2 has produced its first pair.

Let's call it F_5.

F_3 has not produced a pair, because it has been in the cage for only one month.

Total number of pairs in the cage: $F_1 + F_2 + F_3 + F_4 + F_5 = 5$ pairs.

During the fifth month, F_1 produces another pair, as does F_2 (now fully productive). F_3 has been in the cage for a month; so it now also becomes productive, giving birth to its first pair of offspring. The other two rabbit pairs in the cage, F_4 and F_5, are not yet productive. So, at the end of the fifth month, three newborn pairs have been added to the five pairs that were previously in the cage, making the total number of pairs in it: $5 + 3 = 8$:

After the fifth month

F_1 has produced another pair of offspring.

Let's call it F_6.

F_2 has also produced another pair.

Let's call it F_7.

F_3 has produced its first pair of rabbits.

Let's call it F_8.

F_4 and F_5 have not produced pairs, because both have been in the cage for only one month (F_4 was produced by F_1 and F_5 by F_2).

Total number of pairs in the cage: $F_1 + F_2 + F_3 + F_4 + F_5 + F_6 + F_7 + F_8 =$ 8 pairs.

The remainder of the solution proceeds in a similar fashion. It is left as an exercise to the patient reader. By the end of the twelve-month period, 233 pairs will be in the cage. A month-by-month summary of the cumulative rabbit pairs in the cage is given as follows:

After How Long?	How Many Pairs in the Cage?
the start	1 pair
1 month	1 pair
2 months	2 pairs
3 months	3 pairs
4 months	5 pairs

After How Long?	How Many Pairs in the Cage?
5 months	8 pairs
6 months	13 pairs
7 months	21 pairs
8 months	34 pairs
9 months	55 pairs
10 months	89 pairs
11 months	144 pairs
12 months	233 pairs

So, the answer to Fibonacci's puzzle is that there will be 233 pairs in the cage after twelve months. The solution is hardly interesting in itself. But the kinds of surprising patterns it harbors within it are of enormous interest. The first can easily be discerned by putting the successive number of pairs in the cage at the end of each month in a linear sequence, as follows:

1, 1, 2, 3, 5, 8, 13, 21, 34, 55, 89, 144, 233.

Each number in the sequence is equal to the sum of the previous two numbers—for example, 2 (the third number) = 1 + 1 (the sum of the previous two); 3 (the fourth number) = 1 + 2 (the sum of the previous two); and so on. This "hidden formula" in the sequence allows us to extend it, ad infinitum. To get the number after 233, all we have to do is add 233 and 144, which equals 377; to get the one after 377, we add 377 and 233, which equals 610; and so on, ad infinitum:

{1, 1, 2, 3, 5, 8, 13, 21, 34, 55, 89, 144, 233, 377, 610, 987, . . . }.

In mathematics, a sequence or **series** is enclosed by brackets. The numbers in the series are called terms. The three dots indicate that an infinite number of terms follow the ones that are written. The Fibonacci sequence is thus an **infinite series**, in mathematical terminology, meaning it's an ordered succession of numbers or other quantities that goes on ad infinitum. The natural numbers is an example of an infinite series, since there is no last number in it {1, 2, 3, 4, 5, . . . }. If we denote the general term of the series as F_n (for **Fibonacci number**), then the formula that generates each term can be represented as follows:

$$F_n = F_{n-1} + F_{n-2}.$$

This is a shorthand way of indicating that any number in the Fibonacci sequence, F_n, can be determined by adding the one before it, F_{n-1}, to the one before that, F_{n-2}. For readers who may have difficulty in reading such symbols, consider a concrete example. Let's choose $n = 6$. This refers to the "sixth" number in the previous Fibonacci sequence (reading it from left to right). This is the number 8. So, in this case:

$$F_n = F_6 = 8.$$

Consequently, F_{n-1} (the number just before), refers to the fifth number in the sequence. And, as readers can see for themselves, that number is 5:

$$F_{n-1} = F_{6-1} = F_5 = 5.$$

F_{n-2} refers to the fourth number in the sequence. That number is 3:

$$F_{n-2} = F_{6-2} = F_4 = 3.$$

In summary, when $n = 6$, the formula $F_n = F_{n-1} + F_{n-2}$ translates into:

$$F_6 = F_5 + F_4$$
$$8 = 5 + 3.$$

PRIMES VS. COMPOSITES

Primes

An integer (a whole number) is called a **prime number** if its only factors are 1 and itself. A **factor** is a smaller number that divides into a larger number. The larger number is thus made up of smaller factors, which, when multiplied together, produce it. A prime is, in effect, a number that cannot be decomposed into factors. Examples are

$$3 = 3 \times 1$$

$$5 = 5 \times 1$$

$$19 = 19 \times 1$$

Note that the number 1 is not defined as prime.

Composites

An integer is called a **composite number** if it is composed of different factors. Note that in their lowest form, the factors of composite numbers are all prime. Examples include

$$4 = 2 \times 2$$

$$12 = 2 \times 6 = 2 \times (2 \times 3) = 2 \times 2 \times 3$$

$$20 = 2 \times 10 = 2 \times (2 \times 5) = 2 \times 2 \times 5$$

The primes are clearly the "building blocks" of our number system.

Over the years, the properties of the Fibonacci numbers have been extensively studied, resulting in considerable literature. The basic pattern hidden in this series was studied formally by the French-born mathematician Albert Girard (1595?–1632?) in 1632. In 1753, the Scottish mathematician Robert Simson (1687–1768) noted that as the numbers increased in magnitude, the ratio between succeeding numbers approached the Golden Ratio. However, it was the French mathematician and puzzlist Edouard Lucas, whom we will encounter in chapter 6, who detected all kinds of hidden patterns in the Fibonacci numbers and who named it the **Fibonacci sequence**, which is a series of numbers starting with 1 in which each successive number is the sum of the previous two numbers. Lucas also created his own version of the Fibonacci sequence, known as the **Lucas sequence**. It is exactly like the Fibonacci sequence except that it starts with the number 2.

$$\{2, 1, 3, 4, 7, 11, 18, 29, 47, 76, 123, \ldots\}.$$

They have found, for instance, that it provides key insights on prime numbers. In 1962, Vern Emil Hoggart (1921–1981) and Brother Alfred Brousseau (1907–1988) founded a Fibonacci Society for the sole purpose of studying the sequence and the patterns it conceals—and there seem to be an infinitude of them! The society started publication of a periodical, called the *Fibonacci Quarterly*, the year after (1963). It is still being published.

Mathematical Annotations

The number of patterns that can be found in the Fibonacci sequence is mind-boggling. Why would the solution to a simple puzzle contain so many? To the best of my knowledge, no answer exists to that question. All that can be said is that the derivation of the sequence from a simple puzzle gives some substance to the Pythagorean belief that numbers may, after all, constitute the secret language of the universe.

Patterns in the Fibonacci Sequence

Let's take a look at some of the patterns that the Fibonacci sequence conceals. If we take the ratio of two consecutive Fibonacci numbers, it approaches the unending decimal 0.6180339 . . . :

$$\{1, 1, 2, 3, 5, 8, 13, 21, 34, 55, 89, 144, 233, 377, 610, 987, \ldots\}.$$

Ratios of two consecutive numbers in the Fibonacci sequence include:

$$\frac{5}{8} = 0.625$$

$$\frac{8}{13} = 0.615384$$

$$\frac{13}{21} = 0.619047$$

$$\frac{21}{34} = 0.617647$$

...

$$\frac{610}{987} = 0.6180344$$

Etc.

THE GOLDEN RATIO

Also known as the Golden Section, the Golden Ratio is a proportion produced when a line is divided so that the ratio of the length of the longer line segment (**AC**, in the following figure) to the length of the entire line (**AB**) is equal to the ratio of the length of the shorter line segment (**CB**) to the length of the longer line segment (**AC**). The ratio is the unending number 0.6180339 . . .

AC/AB = **CB/AC** = 0.6180339 . . .

A C B

Since antiquity, philosophers, artists, and mathematicians have been intrigued by this ratio, which Renaissance writers called the "divine proportion." It is widely accepted that any form constructed with this ratio exhibits a special beauty. It has also been found to occur mysteriously in nature.

This ratio turns out to be the **Golden Ratio**—a ratio discovered by the Greeks that has been recognized as a standard of aesthetic judgment ever since (see the sidebar). The Fibonacci sequence has so many such patterns in it that one could literally spend a lifetime sifting them out. Here are a few others:

▶ The sequence that results from taking the difference between two consecutive Fibonacci numbers (for example, $3 - 2 = 1$, $8 - 5 = 3$, etc.) produces the original sequence:

{1, 1, 2, 3, 5, 8, 13, 21, 34, 55, 89, 144, 233, 377, 610, 987, . . . }.

Differences between consecutive numbers in the sequence include:

$2 - 1 = 1$
$3 - 2 = 1$
$5 - 3 = 2$
$8 - 5 = 3$
$13 - 8 = 5$
$21 - 13 = 8$
. . . = . . .
↑
Fibonacci sequence (read from the top down)

▶ The sum of the squares of two consecutive Fibonacci numbers is a Fibonacci number. *Note:* if F_n is any Fibonacci number, then F_{n+1} is the number after it. Note also that the symbols F_n^2 and F_{n+1}^2 represent the squares of the consecutive numbers F_n and F_{n+1}:

TABLE 3-1: FIBONACCI NUMBERS

F_n	F_{n+1}	→	F_n^2	+	F_{n+1}^2	=	Fibonacci Number
2	3	→	4 (= 2^2)	+	9 (= 3^2)	=	13
3	5	→	9 (= 3^2)	+	25 (= 5^2)	=	34
5	8	→	25 (= 5^2)	+	64 (= 8^2)	=	89
8	13	→	64 (= 8^2)	+	169 (= 13^2)	=	233
13	21	→	169 (= 13^2)	+	441 (= 21^2)	=	610
21	34	→	441 (= 21^2)	+	1,156 (= 34^2)	=	1,597
. . .	. . .	. . .	. . .	. . .	. . .	. . .	. . .

▶ The third number is 2, and every third number after 2 is a multiple of 2 (= 8, 34, 144, . . .); the fourth number is 3, and every fourth number after 3 is a multiple of 3 (= 21, 144, 987, . . .); the fifth number is 5, and every fifth number after 5 is a multiple of 5 (= 55, 610, . . .); and so on. In general, if the *n*th number in the sequence is *x*, then every *n*th number after *x* turns out to be a multiple of *x*.

One of the most intriguing discoveries related to the Fibonacci sequence is the unexpected relation that it has to **Pascal's triangle**, named after the French philosopher and mathematician Blaise Pascal (1623–1662), a founder of the modern theory of probability. The triangle consists of a triangular arrangement of numbers, whereby a number in a given row is the sum of two numbers immediately above it in the triangle. Here is a section of the triangle:

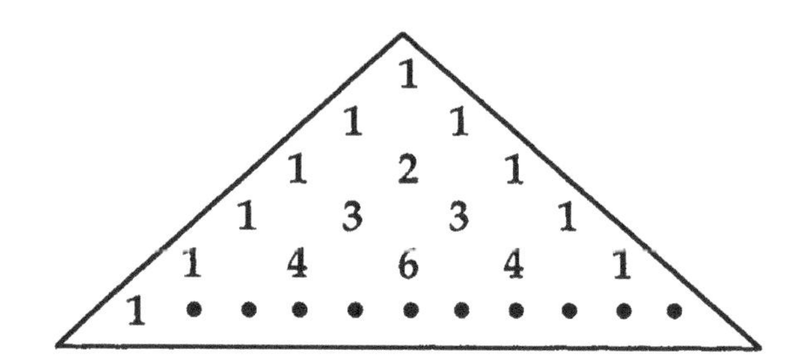

For example, the first 3 in the fourth row from the top is equal to the sum of the two numbers immediately above it (1 + 2). Similarly, the 6 in the fifth row is equal to the sum of the two numbers immediately above it (3 + 3). As it turns out, the diagonal sums of the numbers in Pascal's triangle correspond to the numbers in the Fibonacci sequence {1, 1, 2, 3, 5, 8, 13, 21, 34, . . . }:

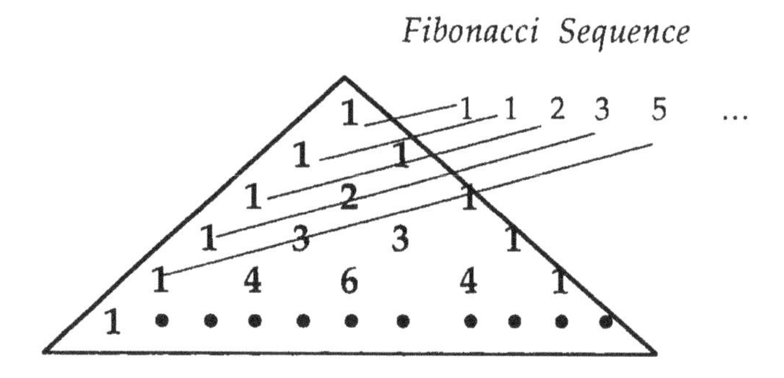

Why do Pascal and Fibonacci numbers correspond in this way? No answer has ever been given to this question, to the best of my knowledge. The correspondence remains a mystery to this day.

Let's now look at a less spectacular, albeit nonetheless fascinating, pattern in the Fibonacci sequence. We start by adding up the first ten consecutive numbers:

1. 1 + 1 + 2 + 3 + 5 + 8 + 13 + 21 + 34 + 55 = 143.

The sum turns out to be divisible by 11 (143 ÷ 11 = 13). Now, amazingly, the same result holds for the sum of any ten consecutive Fibonacci numbers. Take, for example, the ten numbers that start with 55 in the sequence:

2. 55 + 89 + 144 + 233 + 377 + 610 + 987 + 1,597 + 2,584 + 4,181 = 10,857

   and

   10,857 ÷ 11 = 987.

If one examines these two cases more closely, it turns out that the sum of the ten consecutive numbers is equal to 11 times the seventh number in the chosen ten-digit sequence. In example 1, the seventh number is 13, and 13 × 11 = 143; and in example 2, the seventh number is 987, and 987 × 11 = 10,857.

The reader may ask at this point: Why search for such patterns? Do they lead anywhere? Ah, there's the rub, as Shakespeare would have said. Fibonacci numbers seem to rule the universe. There have, in fact, been innumerable applications of Fibonacci numbers to the study of functions, to computer-programming techniques, and to many other areas of mathematics. And, as will be pointed out briefly in the Reflections section, Fibonacci numbers manifest themselves in nature and in all kinds of human affairs. Such remarkable serendipities raise the same question that came to the mind of Paul Dirac (1902–1984), a founder of quantum mechanics, when he was contemplating the discovery that the strength of the electromagnetic force between two electrons yields a constant value of $\frac{1}{137}$. He considered this finding so amazing that he has been quoted as saying that upon arrival to heaven, he would ask God one question only: why $\frac{1}{137}$? To Dirac's question, one could add: "why Fibonacci numbers?"

As mathematicians began to see Fibonacci numbers appear in the most unexpected places and in surprising ways, they grew interested in finding an efficient method to calculate any Fibonacci number. In principle, this is not a problem. To identify the 100th Fibonacci number, for instance, all we have to do is add the 98th and 99th numbers together. However, this still means we have to identify all the numbers up to the 98th, which can prove to be quite tedious. So, in the middle of the nineteenth century, the French mathematician Jacques Binet (1786–1856) elaborated a formula, based on earlier calculations of Leonhard Euler (1707–1856) and Abraham de Moivre (1667–1754). It allows us to find any Fibonacci number, if its position in the sequence, *n*, is known. The Binet formula is given as follows:

$$F_n = \frac{1}{\sqrt{5}}\left[\left(\frac{1+\sqrt{5}}{2}\right)^n - \left(\frac{1-\sqrt{5}}{2}\right)^n\right].$$

It is beyond the objective of this chapter to explain how Binet arrived at the formula. Suffice it to say that it relies entirely on the Golden Ratio. Readers can check the validity of the formula for themselves by plugging various values for n into it.

Sequences and Series

The Fibonacci sequence is, technically, a series—a sequence of numbers, called terms, that is generated by some rule. *Note:* a **negative number** is one that is placed to the left of zero on a number line:

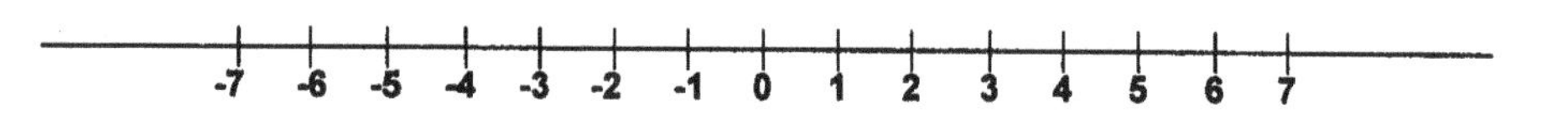

Negative numbers are used, for example, to refer to the weather. In this case, the negative value is said to be "below zero," rather than to "the left of zero" (as shown in the previous figure). This is because the number line on the thermometer is usually read vertically, from top to bottom—not horizontally, from left to right.

NUMBERS

Integers

Integers include the following three groups:

Natural numbers: {1, 2, 3, 4, 5, 6, . . . }

Zero: {0}

Negative numbers: {–1, –2, –3, –4, . . . }

Fractions

Examples of fractions are:

$$\left\{\frac{1}{2}, \frac{-2}{3}, \frac{7}{9}, \frac{14}{23}, \ldots\right\}$$

Integers and fractions together make up the so-called **rational numbers.**

Irrational Numbers

Irrational numbers (also known as radical numbers) are numbers that cannot be expressed as an integer or as a ratio between two integers. Examples include:

$$\{\sqrt{2}, \sqrt{5}, \sqrt{19}, \sqrt{23}, \ldots\}$$

Rational and irrational numbers together make up the *real numbers.*

There are also *transfinite* and *complex numbers* in the real number system. The former will be discussed in chapter 6; the latter are beyond the scope of this book.

The following are examples of series:

1. $-5, -10, -15, -20, -25, \ldots$
2. $5, 10, 20, 40, 80, \ldots$
3. $1, 3, 5, 7, 9, \ldots$
4. $2^2, 2^4, 2^6, 2^8, \ldots$

In series 1, each term differs from the preceding one by –5. In series 2, each term differs from the preceding one by a ratio of 2; that is, each term is produced by multiplying the one before it by 2. In series 3, each term differs from the preceding one by 2. In series 4, the consecutive terms differ by a factor of 2^2; that is, each term is produced by multiplying the previous one by 2^2. Series 1 and 3 are called **arithmetical**, and series 2 and 4 are **geometric**. These will be discussed in chapter 6.

Among the first to study series systematically was the great German mathematician Karl Friedrich Gauss (1777–1855). The story goes that Gauss was only ten years old when he purportedly dazzled his math teacher, after the teacher had asked the class to cast the sum of all the numbers from 1 to 100: $1 + 2 + 3 + 4 + \ldots + 100 = ?$ Gauss raised his hand within seconds, giving the correct response of 5,050. When his teacher asked little Karl how he was able to come up with the answer so quickly, he is said to have replied (more or less) as follows:

> I laid out the numbers in order, removing from the layout the middle number, 50, and the last number, 100: $\{1, 2, 3, \ldots, 97, 98, 99\}$. In this layout, there are forty-nine pairs of numbers that add up to 100. The pairs are made up as follows: the first and last numbers in the layout ($1 + 99 = 100$), the second and second-last numbers ($2 + 98 = 100$), the third and third-last ($3 + 97 = 100$), and so on. That makes 4,900, of course. Adding to this the 50 and 100 that were removed gives 5,050.

In effect, Gauss had discovered and proven how to sum an arithmetical series: $\{1, 2, 3, \ldots 100\}$. The problem given to Gauss and his classmates can be expressed more generally as follows: what is the sum of n consecutive numbers: $\{1 + 2 + 3 + \ldots + n\}$? The answer is: $n(n + 1)/2$. Plugging 100 into n produces the answer:

$$\frac{n(n+1)}{2} = \frac{100(100+1)}{2} = \frac{(100)(101)}{2} = \frac{10{,}100}{2} = 5{,}050.$$

To grasp how the formula was devised, let's put on paper what Gauss did more or less in his head, with a few slight modifications. First, lay out the terms in numerical order and then in reverse order, putting one under the other:

(1)	1	2	3	...	100
	↕	↕	↕		↕
(2)	100	99	98	...	1.

Next, add the two numbers in each column together:

(1)	1	2	3	...	100
+	+	+	+		+
(2)	100	99	98	...	1
Sum	101	101	101	...	101.

Notice that the sum in each case is the same—101. How many times do we get this sum? Since there are 100 columns, that sum occurs 100 times. Thus, the overall sum of the columns is: $100 \times 101 = 10{,}100$. Now, this is twice the sum of all the integers from 1 to 100. Why? Because we added two series together—the top is the initial series in numerical order and the bottom is the same series in reverse order. So, all we have to do is divide the overall sum in half: $10{,}100/2 = 5{,}050$. The previous computations can be summarized arithmetically as follows:

$$\text{Sum of the first 100 numbers} = \frac{100 \times 101}{2} = 5{,}050.$$

Now, let's generalize this arithmetical form. Notice that 101 is 1 more than the number of terms in the series, 100. Thus, if n is the number of terms in a series, $(n + 1)$ stands for 1 more than n:

$$\text{Sum of the first 100 numbers} = \frac{\overset{\overset{n}{\downarrow}}{100} \times \overset{\overset{(n+1)}{\downarrow}}{101}}{2} = 5{,}050.$$

$$\text{Sum of the first } n \text{ numbers} = \frac{n \times (n + 1)}{2}.$$

The formula is written more commonly as $S_{(n)} = n\,(n + 1)/2$. Let's try it out with $n = 15$, which produces the sum of the first fifteen numbers $\{1, 2, 3, 4, + \ldots + 15\}$:

$$S_{(n)} = n\,(n + 1)/2 = 15\,(15 + 1)/2 = (15)\,(16)/2 = 240/2 = 120.$$

We will return to the topic of series in chapter 6. Here, it is sufficient to note that the study of series was made possible in the first place because of the decimal system. As mentioned earlier, its introduction into mathematics was due largely to the efforts of Fibonacci. The Fibonacci sequence itself would hardly reveal its many hidden and intriguing patterns so readily if it were not written with decimal numerals.

Reflections

The apparently infinite numerical patterns that are present in the Fibonacci sequence, which mathematicians continue to extract from it, are not all there is to the story of this truly remarkable series. For some mysterious reason, Fibonacci numbers surface in nature. Daisies tend to have 21, 34, 55, or 89 petals (= the eighth, ninth, tenth, and eleventh numbers in the sequence); trilliums, wild roses, bloodroots, columbines, lilies, and irises also have petals in consecutive Fibonacci numbers. In general, if we start near the bottom of a stalk and count up, the number of sequences of leaves turns out to coincide with some stretch of consecutive numbers within the Fibonacci sequence. And that is not all—Fibonacci numbers also constantly appear in human forms and affairs. A major chord in Western music, for instance, is made up of the third, fifth, and octave tones of the scale—that is, of tones corresponding to the fourth, fifth, and sixth terms in the Fibonacci sequence.

Basically, if one knows how to look, one will find Fibonacci numbers in plants, poems, symphonies, art forms, computers, the solar system, and the stock market. Myriads of books and articles have been written on this topic. It truly boggles the mind to contemplate that all these "unexplained discoveries" can be traced back, ultimately, to a simple puzzle that was designed to illustrate the practicality of the Hindu-Arabic numeral system! The Pythagorean belief that mathematics is the secret language of the cosmos certainly seems to be substantiated by the serendipitous appearance of the Fibonacci sequence in human history. The word *serendipity* was coined by the English writer Horace Walpole (1717–1797) in 1754, from the title of the Persian fairy tale *The Three Princes of Serendip*, whose heroes made many fortunate discoveries accidentally.

Incidentally, the same type of observations can be made about the Golden Ratio, also known by the Greek letter *phi*. The ratio describes the spiraling form of seashells, pine cones, and other symmetries of nature. It is said to have been incorporated by Leonardo da Vinci and Michelangelo into their masterpieces of visual art. And it is apparently found in the proportions used to build the Egyptian pyramids and the Greek Parthenon. As the Pythagoreans suspected, *phi* may provide an important clue to how the universe works, as might the Fibonacci sequence, for that matter.

Explorations

Pattern Detection

26. An infinitude of patterns exists in the Fibonacci sequence. Many are still awaiting discovery. How many patterns can you spot?

27. Let's construct a series in which every number is the sum of the previous three, starting with 1, 2, 3:

$$\{1, 2, 3, 6, 11, 20, 37, 68, 125, 230, 423, 778, \ldots\}.$$

Do you detect any patterns in this sequence?

Miscellaneous Puzzles

Fibonacci's objective was to introduce the decimal number system to a European audience by means of puzzles and practical problems. The following are tricky puzzles that nevertheless bring out the capability of this system to make computation easy.

28. Tim finally came to the realization that smoking is harmful and just plain foolish. So, he decided to quit smoking, as soon as he finished the twenty-seven cigarettes he had left in his pocket. Tim habitually smoked only two-thirds of a cigarette at a time. It was also his habit to reroll the butts into new cigarettes and then smoke them. If he smoked only once each day, how many days went by before he finally quit his bad habit?

29. There are between fifty and sixty people at a party. Jane was counting them, one at a time, when she noticed that if she counted them three at a time, there would be two left over in her calculation method. If, however, she counted them five at a time, there would be four left over. How many people were at the party?

30. There are two containers on a table, A and B. B is twice the size of A. A is half full of wine, and B one-quarter full of wine. Both containers are then filled with water, and the contents are poured into a third container, C. What portion of container C's mixture is wine?

31. During a warehouse fire, a firefighter stood on the middle rung of a ladder, pumping water into the burning warehouse. A minute later, she stepped up three rungs and continued directing water at the building from her new position. A few minutes after that, she stepped down five rungs and, from her new position, continued to pump water into the building.

Half an hour later, she climbed up seven rungs and pumped water from her new position until the fire was extinguished. She then climbed the remaining seven rungs up to the roof of the warehouse to look over the scene from there. How many rungs were on the ladder?

Series

32. As discussed in this chapter, the sum of an arithmetical series is given by the formula: $S_{(n)} = n\ (n + 1)/2$. As we also saw, there is a second main type of series, called a geometric series, which is defined as a series where each successive term differs from the previous one by a constant ratio. For example, in the following series, each term differs from the previous one by a ratio of 2: that is, each term in the series is generated by multiplying the previous one by 2:

$$\{2, 4, 8, 16, 32, 64, 128, \ldots\}.$$

Underlying "structure" of the series:

1st term	2nd term	3rd term	4th term	. . .	*n*th term
2	$4 = 2 \times 2$	$8 = 4 \times 2$	$16 = 8 \times 2$	. . .	?

Can you derive a formula for the general term of this series?

33. What if Gauss's teacher had asked the class to sum up only the even numbers from 1 to 100? Can you figure out a way to do this task quickly? How would you find the sum of all the odd numbers from 1 to 100?

Further Reading

Basin, S. L. "The Fibonacci Sequence as It Appears in Nature." *Fibonacci Quarterly* 1 (1963): 53–64.

Brousseau, Brother A. *An Introduction to Fibonacci's Discovery*. Aurora, S.D.: Fibonacci Association, 1965.

Devlin, Keith. *The Language of Mathematics: Making the Invisible Visible*. New York: W. H. Freeman, 1998.

———. *Mathematics: The Science of Patterns*. New York: Scientific American Library, 1997.

Dunlap, R. A. *The Golden Ratio and Fibonacci Numbers*. Singapore: World Scientific, 1997.

Gardner, Martin. "The Multiple Fascination of the Fibonacci Sequence." *Scientific American* (March 1969): 116–20.

Garland, T. H. *Fascinating Fibonaccis.* White Plains, N.Y.: Dale Seymour, 1987.

Jean, R. V. *Mathematical Approach to Pattern in Plant Growth.* New York: John Wiley & Sons, 1984.

Livio, Mario. *The Golden Ratio: The Story of Phi, the World's Most Astonishing Number.* New York: Broadway Books, 2002.

Pappas, Theoni. *Mathematical Footprints: Discovering Mathematical Impressions All Around Us.* San Carlos, Calif.: World Wide Publishing, 1999.

Vajda, S. *Fibonacci and Lucas Numbers, and the Golden Section.* Chichester, U.K.: Ellis Horwood, 1989.

Vernadore, J. "Pascal's Triangle and Fibonacci Numbers." *Mathematics Teacher* 84 (1991), 314–16.

Vorob'ev, N. N. *Fibonacci Numbers.* New York: Blaisdell, 1961.

Euler's Königsberg Bridges Puzzle

If I were to wish for anything, I should not wish for wealth and power, but for the passionate sense of the potential, for the eye which, ever young and ardent, sees the possible. Pleasure disappoints, possibility never. And what wine is so sparkling, what so fragrant, what so intoxicating, as possibility!

SØREN KIERKEGAARD (1813–1855)

As one of the greatest and most prolific mathematicians of history, Leonhard Euler surely had no time to waste on trivial things. So, the fact that he created puzzles to examine or model mathematical ideas speaks volumes. For example, he devised his famous Thirty-Six Officers Puzzle in 1779 to study the properties of numbers arranged in rows and columns—an idea pattern that led shortly thereafter to the concept of *matrix* in algebra. A **matrix** is defined as an array of numbers or algebraic symbols that can be used for manipulating the numbers or symbols for some specific mathematical purpose such as an arithmetical operation.

However, Euler's most important puzzle is, arguably, his Königsberg Bridges Puzzle, which he formulated in a famous 1736 paper titled "The Seven Bridges of Königsberg." He no doubt suspected that the puzzle bore implications for mathematics. But even he could not possibly have imagined that it contained so many revolutionary insights—insights that would eventually lead to the establishment of two autonomous branches, known today as graph theory and topology. For this reason, and because it never fails to intrigue people who come across it for the first time, Euler's puzzle ranks, clearly, as one of the top ten of all time.

LEONHARD EULER (1707–1783)

Euler was born in Basel, Switzerland. He studied under Johann Bernoulli (1667–1748), an important figure in the development of the calculus. From 1727 to 1766, he worked as a professor of both mathematics and physics at institutions in St. Petersburg and Berlin. Euler's major contributions were to number theory, a field that he helped to found.

In his *Introduction to the Analysis of Infinities* (1748), Euler gave the first full treatment of the basic principles and the methods of algebra, trigonometry, and analytical geometry. Although chiefly a mathematician, he also made contributions to astronomy, mechanics, optics, and acoustics.

The Puzzle

In the German town of Königsberg runs the Pregel River. In the river are two islands, which in Euler's time were connected with the mainland and with each other by seven bridges. The residents of the town often debated whether it was possible to take a walk from any point in the town, cross each bridge once and only once, and return to the starting point. No one had found a way to do it, but, on the other hand, no one could explain why it seemed to be impossible. Euler became intrigued by the debate, turning it into one of the greatest puzzles of all time:

> In the town of Königsberg, is it possible to cross each of its seven bridges over the Pregel River, which connect two islands and the mainland, without crossing over any bridge twice?

In the following schematic map of the area, the land regions are represented by capital letters (**A, B, C, D**) and the bridges by lower-case letters (*a, b, c, d, e, f, g*):

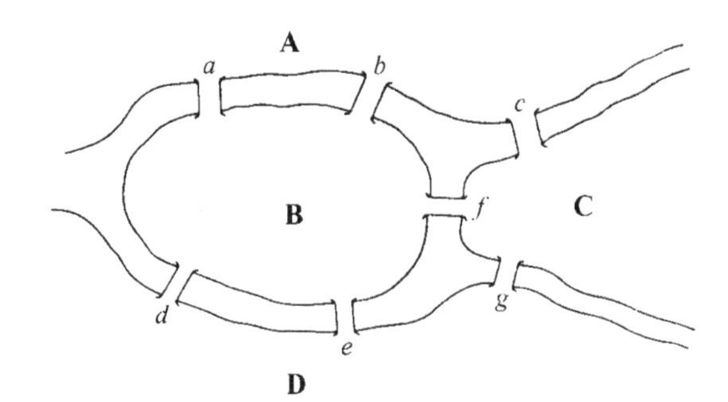

Euler went on to prove that it is impossible to trace a path over the bridges without crossing at least one of them twice. He started by reducing the map of the area to outline form, known as a *graph*, and restating the puzzle as follows:

> Is it possible to draw the following graph without lifting pencil from paper and without tracing any edge twice?

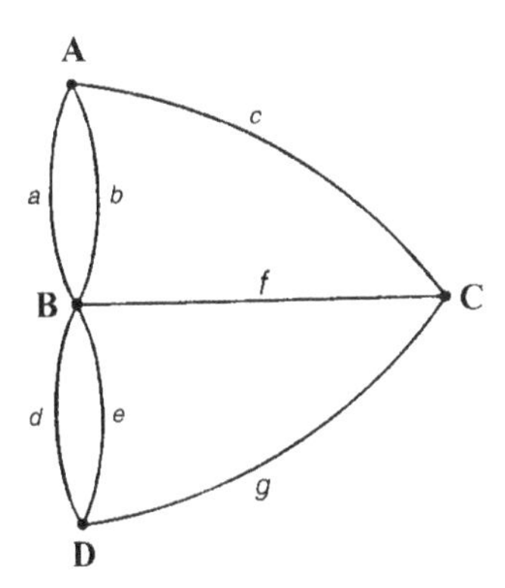

As Euler realized, the graph version of the puzzle provides a more manageable depiction of the situation because it disregards the distracting shapes of the land masses and the bridges, reducing them to *points* or *vertices* and portraying the bridges as *paths* or *edges*. This is called a *network* in contemporary graph theory.

In order to grasp Euler's solution, it is helpful to look at a few simple networks with different even and odd vertices. An *even vertex* is one where an even number of paths converge, and an *odd vertex* is one where an odd number of paths converge.

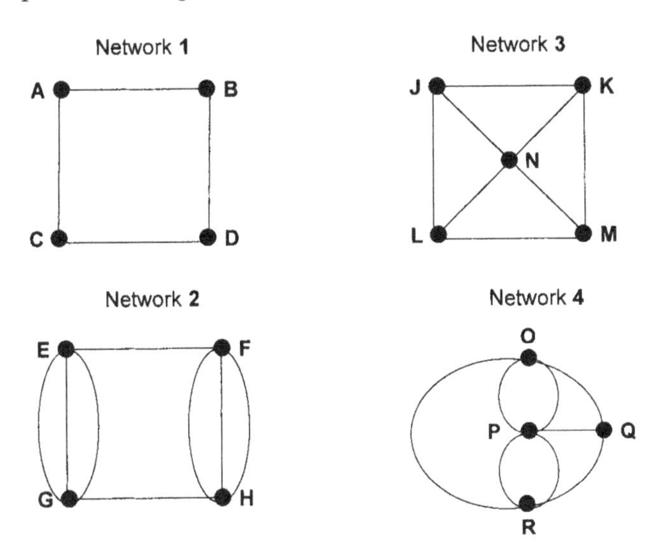

In network 1, an even number of linear paths (two) converge at each of its four vertices. Starting at any vertex, this network can be easily traversed without having to double back over any path. In network 2, an even number of paths (four—two linear and two curved) converge at each of its four vertices. Again, by tracing the network with a pencil, the reader will be able to see that this one, too, can be easily traversed, without having to double back over a path already traced. In network 3, an odd number of linear paths (three) meet at each of the four outer vertices (**J**, **K**, **L**, **M**), and an even number (four) meet at the inner vertex, **N**. Tracing this network, without doubling back, turns out to be impossible. In network **4**, the top vertex, **O**, is even, with four curved paths converging there; the one right below it, **P**, is odd, with one linear and four curved paths meeting there; the bottom vertex, **R**, is even, with four curved paths converging there; and the vertex **Q**, to the right, is odd, with two curved paths and one linear path meeting there. In total, there are two odd and two even vertices. Network 4 can be traversed without doubling back over any path (as readers can verify for themselves).

Is there any hidden pattern here? Creating more complex networks, with more and more paths and vertices in them, will show that it is not possible to traverse a network that has more than two odd vertices in it without having to double back over some of its paths. Euler proved this very fact in a remarkably simple way:

▶ A network can have any number of even paths in it, because all the paths that converge at an even vertex are "used up" without having to double back on any one of them. For example, at a vertex with just two paths, one path is used to get to the vertex and another one to leave it. Both paths are thus used up without our having to double back over either one of them. Take, as another example, a vertex with four paths: one path gets us to the vertex, and a second one gets us out. Then, a third path brings us back to the vertex, and a fourth one gets us out. All paths are once again used up. The same reasoning applies to any even vertex.

▶ At an odd vertex, on the other hand, there will always be one path that is not used up. For example, at a vertex with three paths, one path is used to get to the vertex and another one to leave it. But the third path can only be used to go back to the vertex. To get out, we must double back over one of the three paths. The same reasoning applies to any odd vertex.

▶ Therefore, a network can have, at most, two odd vertices in it. And these must be the starting and ending vertices. Why? Let's label one odd vertex **A** and the other **B**. Being an odd vertex, at **A** there will be one path not used up. Similarly, at **B** there will also be one path not used up. However, if one of these paths is used to start off and the other to get us to the end, the two will be, in effect, used up.

▶ If there is any other odd vertex in the network, however, there will be a path or paths over which we will have to double back.

Now, let's apply this principle to the previous Königsberg graph. The network has four vertices in it. Each one is odd, as readers can confirm for themselves: **A** = 3, **B** = 5, **C** = 3, **D** = 3. This means that the network cannot be traced by one continuous stroke of a pencil without having to double back over paths that have already been traced. So, with his ingenious proof, Euler resolved the debate over the Königsberg bridges, once and for all.

Mathematical Annotations

It is impossible to deal in any detailed way with the implications of Euler's puzzle for modern graph theory, topology, and the mathematical study of impossibility. That would take a huge volume, in itself. The discussion here will thus be limited to an examination of basic notions in these areas.

Graph Theory and Topology

Graph theory has had a great impact on mathematical method, bringing together areas that were previously thought to be separate. **Graph theory** is now a branch of mathematics that deals with the description of graphs of all kinds. A **graph** is any diagram consisting of nodes (also known as vertices) that can or cannot be connected by lines (also known as edges). Higher-dimensional graphs are called **planar** and **nonplanar**. A path that traverses every edge of a graph exactly once is called **Eulerian**. The path **D-E-A-B-C-F-E-B-F-D-C** in the following graph is Eulerian, as readers can confirm for themselves:

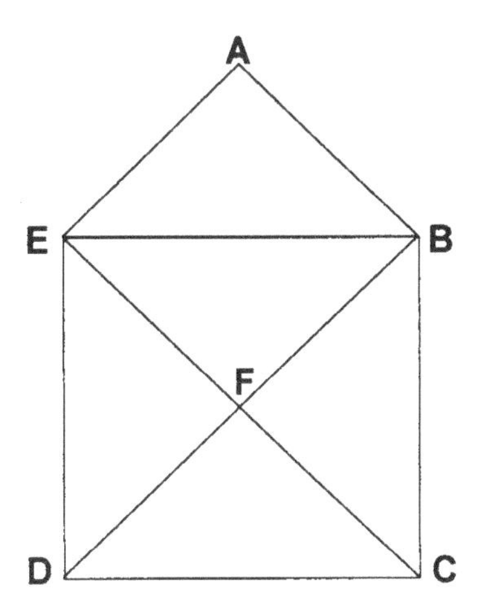

A well-known example of a graph that may or may not be Eulerian is the **Hamiltonian circuit**, named after the Irish mathematician William Rowan Hamilton (1805–1865). Formally, it is a path traced on a graph that visits each vertex on the graph only once, except possibly for the start and the finish, which may be on the same vertex. Hamilton presented it as a game in 1857 called *Around the World*. The object of the game is to travel around the world, along the edges of the following map, visiting each of the twenty cities exactly once:

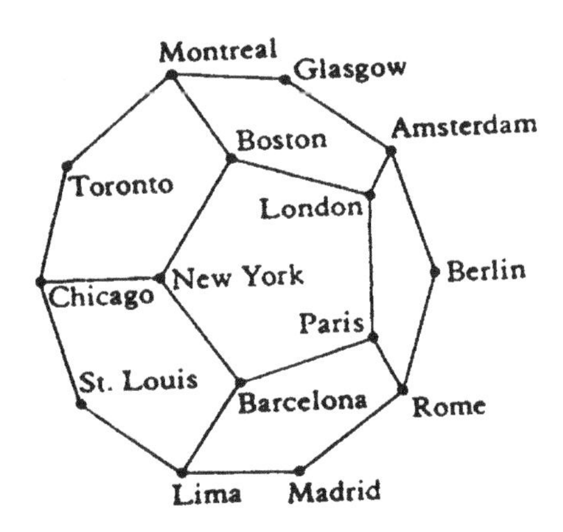

Hamiltonian circuits must be examined individually, and finding a Eulerian path—if one exists—is a matter of trial and error, insight, and luck! Readers may wish to play Hamilton's game, finding out for themselves if the map can be traversed as stipulated.

Several decades after Euler's proof of the Königsberg problem, mathematicians began to study figures that retained their structural features after being deformed. The observation of such figures led, over time, to the study of shapes and their properties, developing gradually into an independent branch of mathematics called topology. The first comprehensive treatment of the field, titled *Theory of Elementary Relationships*, was published in 1863. It was written by the German mathematician Augustus Möbius (1790–1868), the inventor of a truly enigmatic figure called the **Möbius strip**.

To make such a figure, start by drawing a dotted line down the middle of a flat rectangular strip of paper. Give the strip a half-twist (through 180 degrees) and then join the ends. Readers are invited to carry out these instructions to create such a strip for themselves:

Now, how many sides does this strip have? Running a pencil along the dotted line brings us right back to where we started:

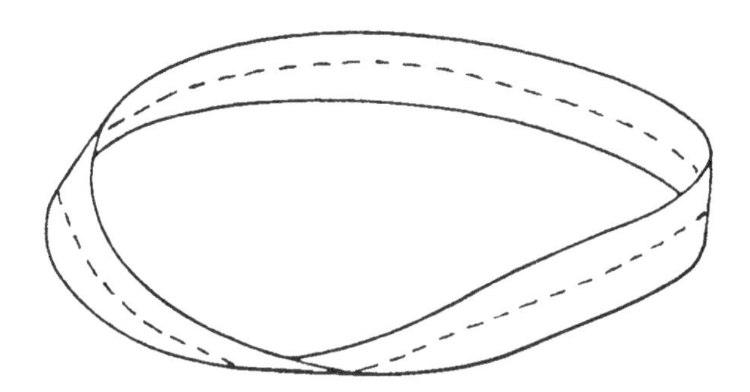

It would appear that the strip has only one side, although the original unjoined strip had two! Even more perplexing is the fact that if one cuts the Möbius strip in two along the pencil line, it does not come apart. As if by magic, two strips linked together into one are produced—as readers can verify for themselves. That strip, made of the two strips, is twice as long and half as wide as the original strip!

The German mathematician Felix Klein (1849–1925) became so captivated by the Möbius strip that in 1882 he invented a "bottle version" of it—known appropriately as the **Klein bottle**:

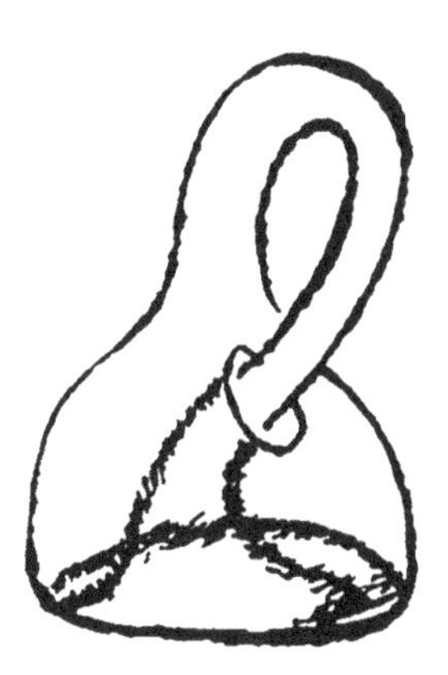

The bottle is a one-sided closed shape with no ends. Yet it has no inside! Indeed, if water were poured into it, the water would come out of the same hole into which it was poured. If cut in two lengthwise, the bottle forms two Möbius strips. Now, how could Klein have made his sense-defying bottle? Its basic construction principle is actually quite straightforward. Take a rubber tube and then puncture a hole in it so that one end can be inserted within it as shown:

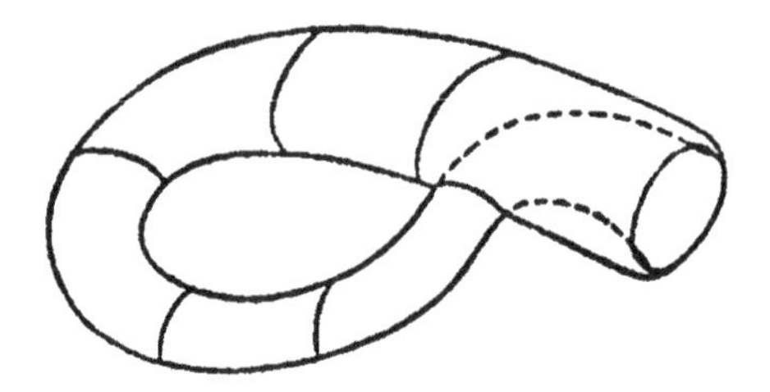

The resulting surface is a closed one with no break. If we start a path anywhere that cuts the surface at any point, we will end up at the point from which we started. No matter where we penetrate the surface, we are still outside of it.

Now, the reader may ask: as intriguing as they are, of what use are such topological oddities? It would take a large tome to answer this question. Suffice it to say here that not only have such bizarre shapes been important to the development of topology, they have also had many applications and implications—conveyor belts and audio tapes designed as Möbius strips wear equally on both sides and thus can be used for much longer periods of time; DNA has a Möbius structure; the universe, too, is thought by many scientists to have this very structure, and the list could go on and on.

Topology concerns itself with determining such things as the *insideness* or the *outsideness* of shapes. A circle, for instance, divides a flat plane into two regions, an inside and an outside. A point outside the circle cannot be connected to a point inside it by a continuous path in the plane without crossing the circle's circumference. If the plane is deformed, it may no longer be flat or smooth, and the circle may become a crinkly curve, but it will continue to divide the surface into an inside and an outside. That is its defining structural feature. Topologists study all kinds of figures. They investigate, for example, knots that can be twisted, stretched, or otherwise deformed, but not torn. Two knots are equivalent if one can be deformed into the other; otherwise, they are distinct.

Euler himself discovered several fundamental topological properties about figures. In the case of a three-dimensional figure, for instance, he found that if we subtract the number of edges (e) from the number of vertices (v) and then add the number of faces (f), we will always get 2 as a result:

$$v - e + f = 2.$$

Take, for example, a cube:

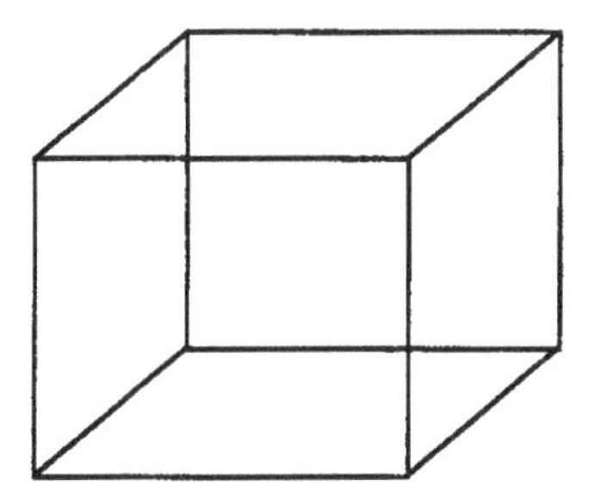

How many vertices (sharp corners) does it have? The answer is eight. How many edges does it have? The answer is twelve. How many faces (flat sides) does it have? The answer is six. Now, inserting these values in the formula, you can see that the relation it stipulates holds:

$$v - e + f = 2$$
$$8 - 12 + 6 = 2.$$

Now, let's try out this formula on a tetrahedron (pyramid):

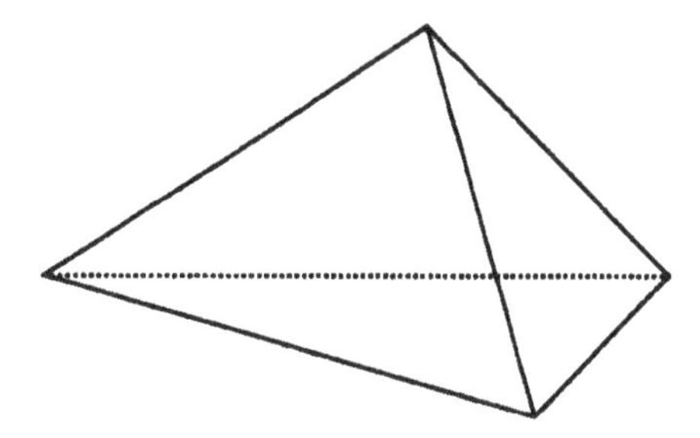

As can be seen, there are four vertices, six edges, and four faces in this case. Thus:

$$v - e + f = 2$$
$$4 - 6 + 4 = 2.$$

Euler also proved that for plane figures, the value of $v - e + f$ is 1, rather than 2. A rectangle, for instance, has four vertices, four edges, and one face:

Therefore:

$$v - e + f = 1$$
$$4 - 4 + 1 = 1.$$

The Königsberg Bridges graph, being a planar graph, also possesses this property. It has four vertices, seven edges, and four faces. Therefore:

$$v - e + f = 1$$
$$4 - 7 + 4 = 1.$$

Take, as one last example, the following graph:

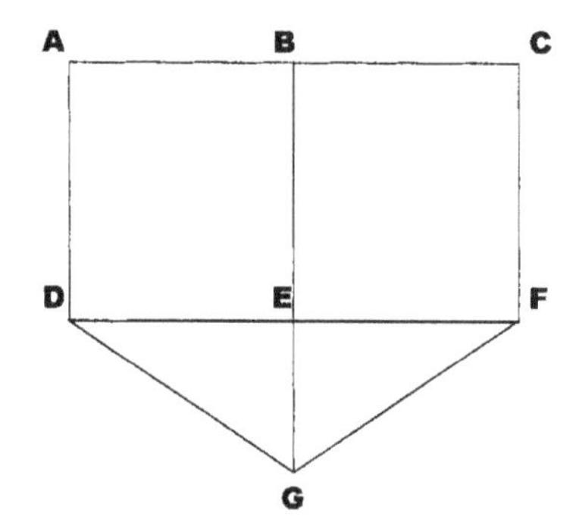

The number of vertices (**A, B, C, D, E, F, G**) is seven, the number of edges (**AD, DE, AB, BE, BC, CF, EF, FG, EG, DG**) is ten, and the number of faces is four (rectangles **ADEB, BEFC**, and triangles **DEG, EFG**). Thus:

$$v - e + f = 1$$
$$7 - 10 + 4 = 1.$$

Euler proved this relation with a remarkably simple procedure. Take, for example, the following rectangle with a diagonal in it. In graph terms: $v = 4$, $e = 5$, and $f = 2$:

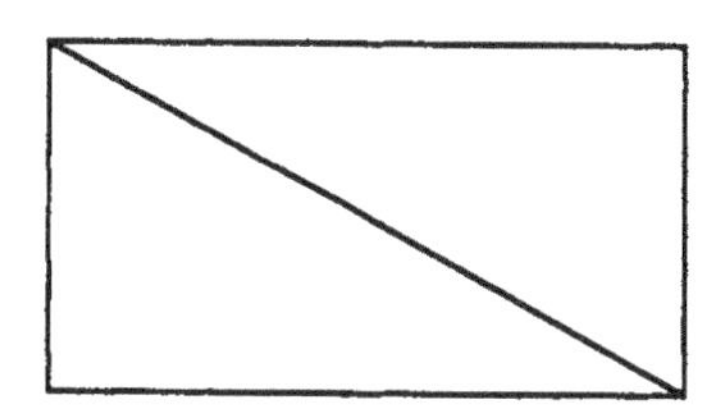

Thus:

$$v - e + f = 1$$
$$4 - 5 + 2 = 1.$$

If we remove the diagonal, which is an edge, we also decrease the number of faces by one, because the graph becomes a rectangle. Since the number of vertices remains unchanged, the relation holds:

$$v - e + f = 1$$
$$4 - 4 + 1 = 1.$$

In general, if we remove an edge from a graph, we are simultaneously removing a face from it. This leaves the value of the relation unaltered. Now, if we eliminate a vertex, we are also removing the edge that goes into it, of course. This reduces v and e by one but leaves the formula again unchanged in value.

Impossibility

The Königsberg Bridges Puzzle not only provided the basic insights that led to the establishment of two new branches of mathematics—graph theory and topology—but also held significant implications for the study of mathematical impossibility. Euler's demonstration that the Königsberg network was impossible to trace without having to double back on at least one of the paths showed how the question of impossibility can be approached systematically.

As another example of how something can be shown to be impossible, consider the following problem:

Find five consecutive odd numbers that add up to 64.

Let's start by considering the sum of the first five odd numbers in sequence:

$$1 + 3 + 5 + 7 + 9 = 25.$$

If we continue adding sets of five consecutive odd numbers, we will find that the sum turns out to be constantly odd—as readers can confirm for themselves. It would seem, therefore, that it is impossible for five consecutive odd numbers to add up to an even sum, such as 64.

Is there any way to prove this? As discussed in the answer to Exploration number 11 (chapter 1), the formula $(2n + 1)$ stands for any odd whole number. Since two consecutive odd numbers differ by 2—for example,

1 and 3 differ by 2, 5 and 7 also differ by 2, and so on—then if the first one in a sequence of five consecutive odd numbers is represented by $(2n + 1)$, the one after it can be represented with the expression $(2n + 3)$, the third with $(2n + 5)$, the fourth with $(2n + 7)$, and the fifth with $(2n + 9)$. Adding up these five consecutive odd numbers yields the following result:

$$(2n + 1) + (2n + 3) + (2n + 5) + (2n + 7) + (2n + 9) = (10n + 25).$$

Now, consider the expression $(10n + 25)$. In it, the term $10n$ is a number ending in 0 because any number n multiplied by 10 will invariably produce a digit ending in 0: $1 \times 10 = 10$, $2 \times 10 = 20$, $15 \times 10 = 150$, and so on. The second term in the expression is 25. It is to be added to the previous digit ending in 0. This means that the result will always end in the digit 5: $10 + 25 = 35$, $20 + 25 = 45$, $150 + 25 = 175$, and so on. So, the expression $(10n + 25)$ represents an odd digit, no matter what n is.

The ancient Greeks grappled constantly with the concept of impossibility, wondering why, for example, it was seemingly impossible to trisect an angle with a compass and a ruler, given that bisection was such a simple procedure. To bisect an angle, such as the following figure, ∠**AOC**, place the compass on point **O** and draw an arc that intersects the sides of the angle at points **X** and **Y**. Then extend the width of the compass to a length greater than half the distance from **X** to **Y**. Now, place the compass on **X**, and draw an arc in the interior of ∠**AOC**. Repeat the last procedure with the compass on **Y**. Label the point of intersection **P**. Finally, draw the line **OP**. This line bisects ∠**AOC**:

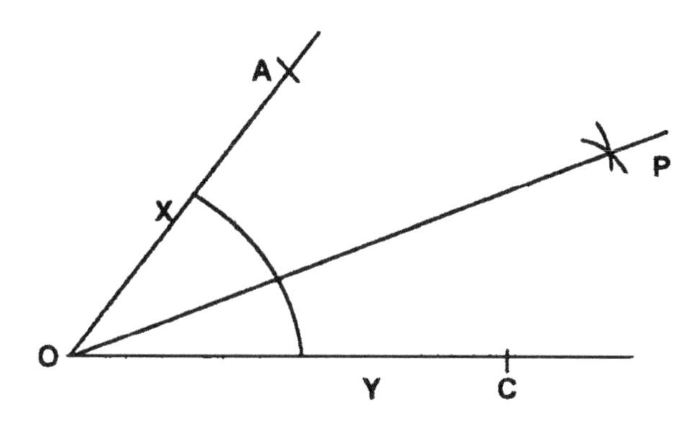

For years, mathematicians attempted trisection with a compass and a ruler, but always to no avail. The demonstration that it was impossible had to await the development and the spread of Descartes' method of converting every problem in geometry into a problem in algebra. The proof that trisection was impossible was based on this method. It came in the nineteenth century, after mathematicians had established that the equation that corresponds to trisection must be of degree 3—that is, it must be an equation in which one of its variables is to the power of 3: for example, $x^3 - 2x^2 + x = 0$. A construction carried out with a compass and a ruler translates, on the other hand, into an equation to the second degree: for example, $x^2 - 14 = 0$. Thus, trisection with a compass and a ruler is impossible. The formal proof was published by the mathematician Pierre Laurent Wantzel (1814–1848) in 1837.

To conclude the discussion of impossibility, I cannot help but mention the Fifteen Puzzle—devised in 1878 by none other than Sam Loyd, one of the cleverest puzzlists of all time (whom we will meet in chapter 7). As a mass-produced gadget, it became a craze that swept across America and Europe. Employers in many states even put up notices that prohibited playing the game during office hours. In France, it was decried as a greater scourge than alcohol or tobacco. It is still popular and is being sold throughout the world.

Loyd put fifteen consecutively numbered sliding blocks in a square plastic tray large enough to hold sixteen such blocks. The blocks are arranged in numerical sequence, except for the last two, fourteen and fifteen, which are in reverse order. The object of the puzzle is to arrange the blocks into numerical sequence from 1 to 15, by sliding them, one at a time, into an empty square, without lifting any block out of the frame:

The Fifteen Puzzle

1	2	3	4
5	6	7	8
9	10	11	12
13	15	14	

The puzzle, as it turns out, is impossible to solve, but it made the wily Loyd a considerable amount of money nonetheless. People simply cannot ignore a challenge, no matter what the costs are in time and energy. Incidentally, Loyd offered a prize of $1,000 for the first correct solution, knowing full well that the puzzle could never be solved.

Notice that when the blocks are in numerical order, each one is followed by a block that is exactly one digit higher (1 is followed by 2, 2 is followed by 3, and so on):

1	2	3	4
5	6	7	8
9	10	11	12
13	14	15	

In any other arrangement, some blocks will be followed by blocks that are numerically lower (for example, 2 followed by 1, 4 followed by 3, etc.). Every instance of a block followed by one that is lower than itself can be called an inversion. If the sum of all the inversions in a given arrangement is even, a solution is possible. If the sum is odd, it is impossible. For instance, the following sequence of blocks can be rearranged into numerical order because the sum of the values of the inversions is 6—an even number (2 is followed by 1, 4 is followed by 3, 6 is followed by 5, 8 is followed by 7, 10 is followed by 9, 12 is followed by 11):

2	1	4	3
6	5	8	7
10	9	12	11
13	14	15	

Loyd's game has only one inversion (15 is followed by 14). This is an odd number, and thus it is impossible to rearrange the blocks in numerical order.

Reflections

Graph theory and topology did not come into their own as branches of mathematics until the middle part of the nineteenth century, but unquestionably, their foundation was laid by Euler's puzzle. And, as we saw in this chapter, the method Euler used to solve his puzzle also laid the groundwork for systematically investigating the mathematical notion of "impossibility."

All of this reveals, in microcosm, how mathematical progress unfolds. At first, an insight comes (generally, from a puzzle). This then leads to the development of a series of conjectures, which become theorems once they are proved. The proofs are based on definitions, previously proved theorems, and logical reasoning. These subsequently enable mathematicians not only to understand the original insight better, but also to see relationships among ideas and facts that were previously considered to be separate or unrelated but that, in effect, turn out to have a common structure.

Explorations

Graphs and Networks

34. Let's start with a few simple exploratory exercises. The following two graphs are Eulerian. Find a Eulerian path through each one:

A.

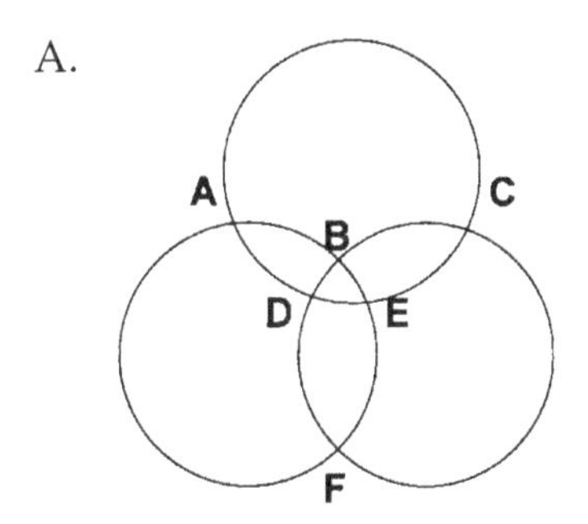

B.

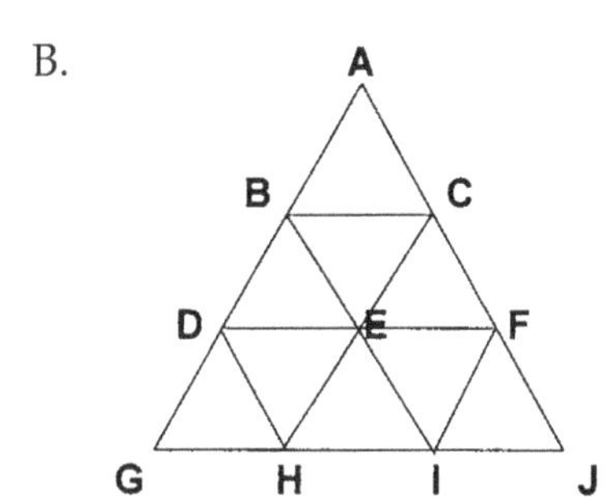

35. Indicate which of the following paths are Eulerian and which are not:

A.

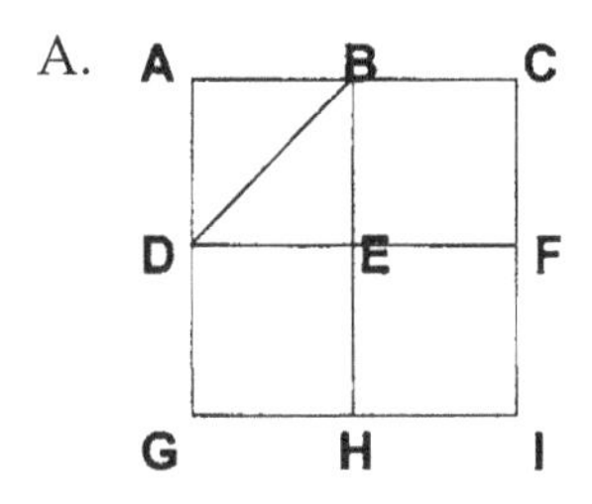

B.

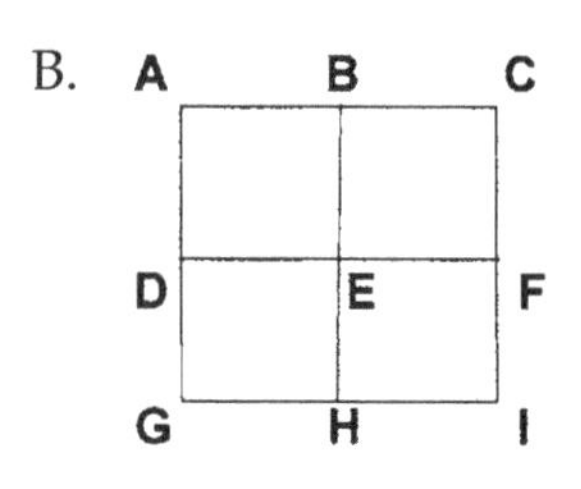

C.

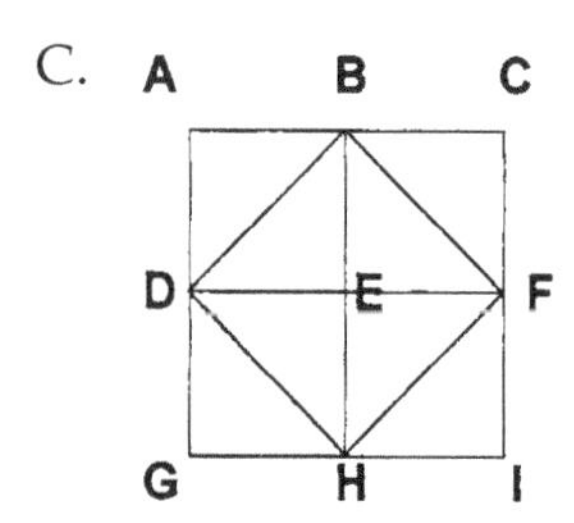

D.

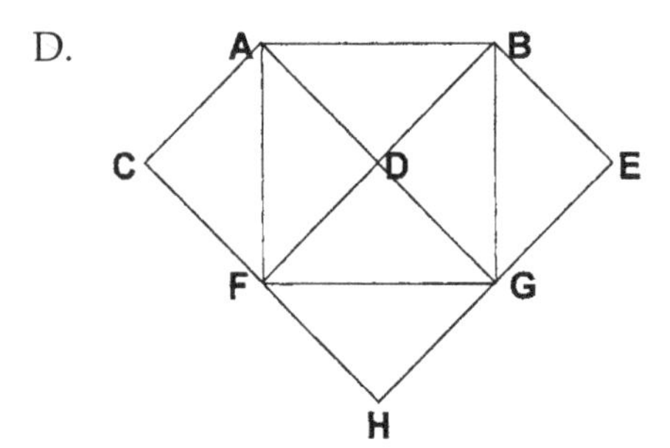

36. Construct any two graphs: a Eulerian and a non-Eulerian one. This will allow you to explore the basic themes of this chapter, directly and creatively.

37. Test the validity of the Eulerian relationship $v - e + f = 2$ on an octahedron (an eight-sided three-dimensional figure):

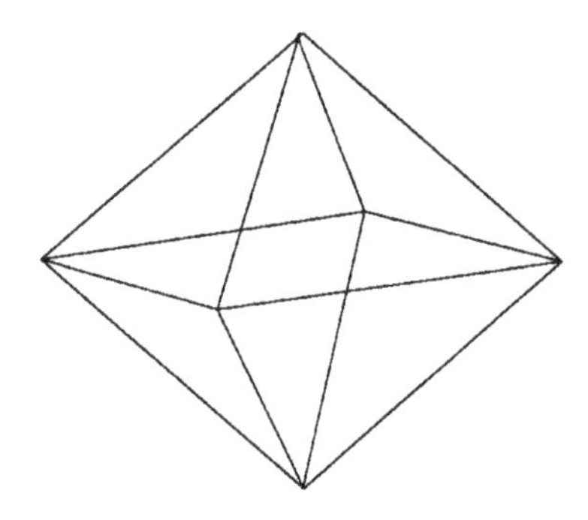

38. Test the validity of the Eulerian relationship $v - e + f = 1$ on the following plane figures:

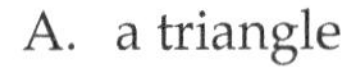

A. a triangle

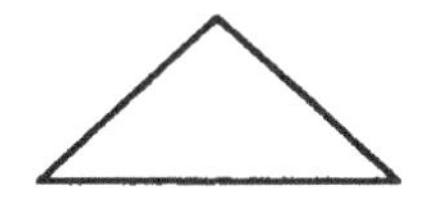

B. a square

C. a pentagon

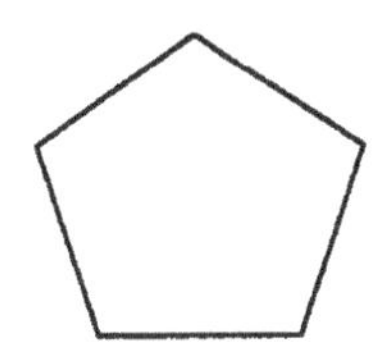

D. a hexagon

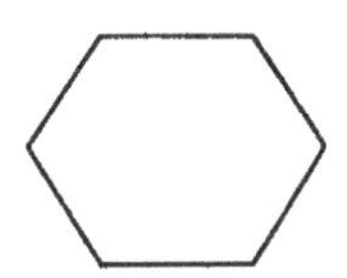

Do you detect any pattern?

39. A rectangle is a Eulerian graph, since it can be traversed once without our having to retrace any of its edges (sides):

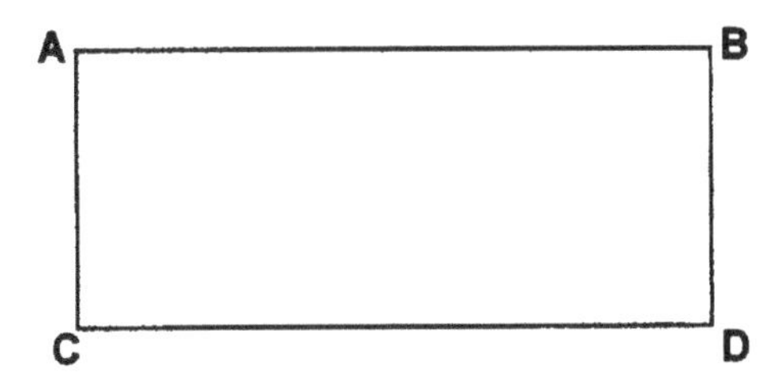

One Eulerian path is **A-C-D-B-A**. Now, if we add diagonals to the rectangle, we would produce a non-Eulerian graph, because each of the four vertices becomes odd—since now three edges converge at each one. The only even vertex would be the point of intersection, **E**, of the diagonals—where four edges converge:

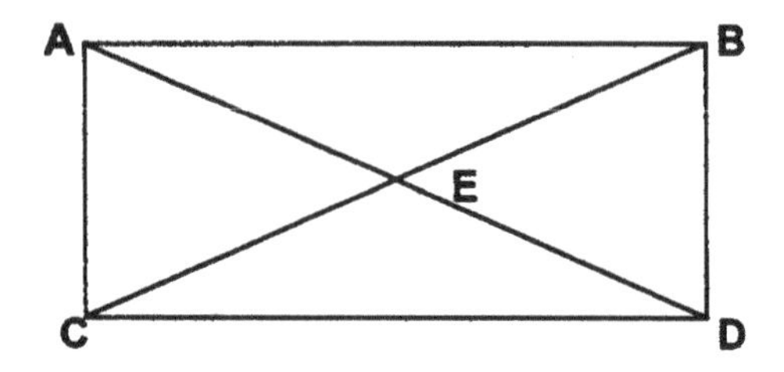

Can you transform the above graph into a Eulerian one?

Impossibility

40. The reader is encouraged to make a model of Loyd's puzzle with simple equipment and then arrange the blocks in such a way as to make the puzzle solvable. Here is one possible "solvable" arrangement:

2	1	4	3
5	6	7	8
9	10	11	12
13	14	15	

41. Can you find two consecutive odd numbers whose product is 316?

Further Reading

Adler, Irving. *Monkey Business.* New York: John Day, 1957.

Falletta, Nicholas. *The Paradoxicon: A Collection of Contradictory Challenges, Problematic Puzzles, and Impossible Illustrations.* New York: John Wiley & Sons, 1983.

Penrose, L. S., and R. Penrose. "Impossible Objects: A Special Type of Visual Illusion." *British Journal of Psychology* 49 (1958): 31–33.

Guthrie's Four-Color Problem

There never comes a point where a theory can be said to be true. The most that one can claim for any theory is that it has shared the successes of all its rivals and that it has passed at least one test which they have failed.

A. J. AYER (1910–1989)

FROM THE DAWN OF HISTORY, people have made maps to help them locate places, measure distances, plan trips, and in general to find their way around the earth. Pilots of ships have used maps to navigate and explore the world.

The art of map making requires, above all else, accuracy in representation. This often involves the use of different tints to make regions on maps visibly unique. It is, in fact, from coloring maps distinctively that one of the greatest puzzles of all time came into being in the middle part of the nineteenth century. Already in antiquity, map makers noticed that four tints seemed to be sufficient to color any map, so that no two contiguous (touching) regions would share a color. For example, the following eight states, which cluster around each other as shown, can be colored distinctively with just four colors, as follows: (1) one color for Illinois, Tennessee, and Virginia, (2) a second color for Missouri and Ohio, (3) a third for Indiana and West Virginia, and (4) a fourth for Kentucky. In this way, no two bordering states share the same color:

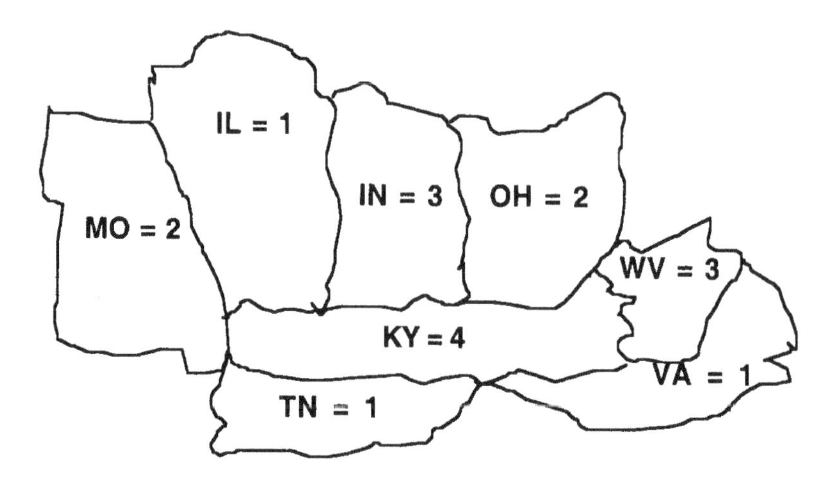

In 1852, a young mathematician at University College, London, named Francis Guthrie (1831–1899), was coloring maps when he realized that four tints were seemingly sufficient to color any map in such a way that adjacent regions (that is, those sharing a common boundary segment, not just a point) would show different colors. Not able to figure out a way to prove this himself, he asked his brother Frederick if he knew of any principle or theorem that proved it. Frederick passed his sibling's query on to the famous mathematician Augustus De Morgan (1806–1871), who, unaware of any existing proof, immediately grasped the mathematical implications that Guthrie's question held. Word of the problem spread quickly. Thus was born the Four-Color Problem.

It is not a puzzle in the traditional sense, since it arose initially from the observations of map makers. Nevertheless, it has all the structural features of a true puzzle. Indeed, it required a large dose of insight thinking to solve. However, although seemingly solved, its demonstration leaves many mathematicians wary. Proof of the "Four-Color theorem" is so different from traditional proofs that it has, in fact, led to a debate on the foundations of mathematical method. For this reason, the Four-Color Problem belongs on the list of the top ten puzzles of all time.

The Puzzle

Although there is evidence that Augustus Möbius (whom we met in the previous chapter) had discussed the Four-Color Problem in a lecture to his students in 1840, Guthrie's version, as related to De Morgan, made the problem famous. In its simplest form, the Four-Color Problem reads as follows:

> What is the minimum number of tints needed to color the regions of any map distinctively? (If two regions touch at a single point, the point is not considered a common border.)

It may be useful to provide a slightly different formulation of the same problem, for the sake of clarity:

> What is the least possible number of colors needed to fill in any map, so that neighboring countries are always colored differently?

To discuss the essential quality of the problem and the challenges it poses, it is useful to start by looking at specific cases. In map 1, there are two touching regions (that is, regions that have a common border); and in map 2, there are three touching regions. In the first map, two colors are needed to keep the regions distinct; and in the second, three are required. Note that numbers are used to represent colors:

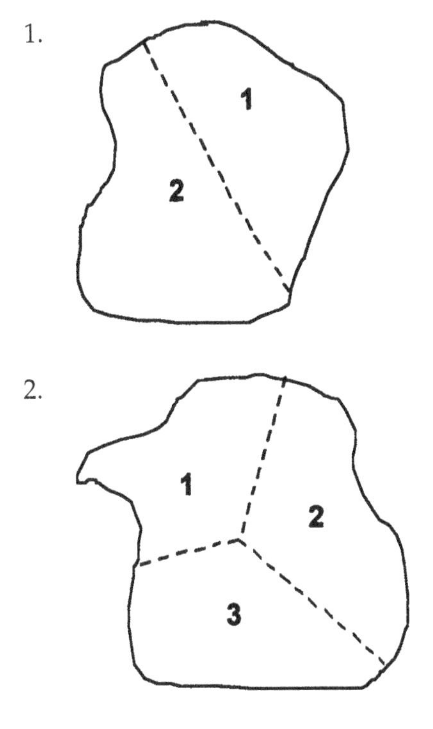

In the next map, there are four regions. Each region shares boundaries with each of the other three: 1 touches 2, 3, and 4; 2 touches 1, 3, and 4; 3 touches 1, 2, and 4; and 4 touches 1, 2, and 3. As shown, four colors (1, 2, 3, 4) will do the job of ensuring that no two regions share a boundary of the same color:

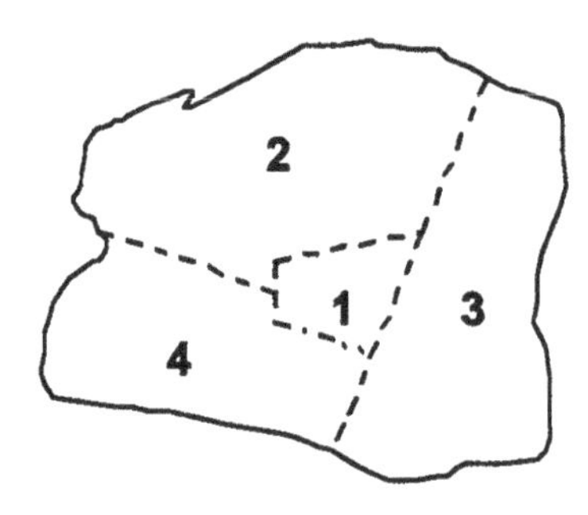

The following map has five touching regions. How many colors are required in this case to make sure that no two regions share a boundary of the same color? As can be seen, three colors (1, 2, 3) are sufficient to do the job:

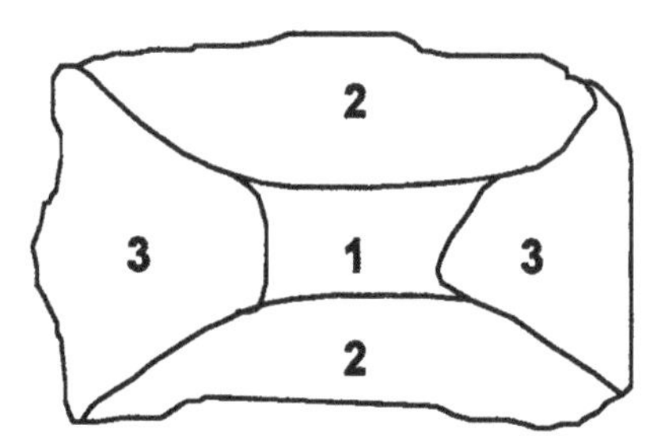

Finally, the next map has nineteen regions. As can be seen, four colors (1, 2, 3, 4) are once again sufficient to color it distinctively:

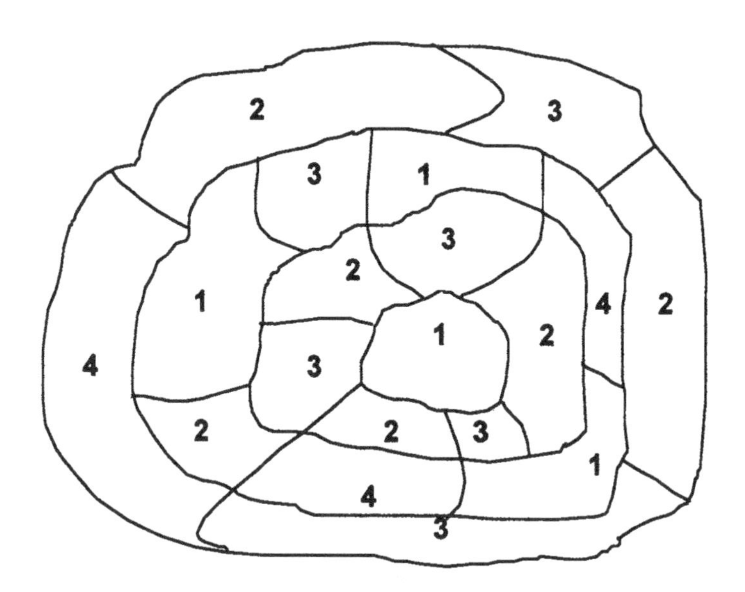

One might think that as maps become more and more complicated, with more regions, an increasing number of colors will be required to differentiate the regions. But as the example of the nineteen-region map suggests, four colors always seem to do the job nicely. The challenge is to prove this very supposition—namely, that four colors are *sufficient* to color any map, no matter how many regions it has.

After De Morgan made the Four-Color conjecture widely known, mathematicians started in earnest trying to prove it with the traditional "Euclidean" methods of proof. But their efforts consistently proved to be fruitless.

EUCLID (c. 330–270 B.C.)

Among the first to establish mathematics as a theoretical enterprise based on methods of proof was the Greek mathematician Euclid. In his textbook called the *Elements,* Euclid began with accepted or self-evident truths called **axioms** (for example, two straight lines intersect at only one point, not two or more) and **postulates** (a statement that requires no proof). From them, he derived theorems.

The place and the date of Euclid's birth are uncertain. It is known that he taught mathematics at the Museum, an institute in Alexandria, Egypt. Euclid probably studied in Athens and came to Alexandria in 300 B.C. at the invitation of the Egyptian ruler Ptolemy I. It is said that when Ptolemy asked him whether there was a shorter way to learn geometry than the *Elements,* Euclid replied ironically, "There is no royal road to geometry."

We encountered two of these methods in chapter 1, when we proved that the vertically opposite angles that are formed when two straight lines intersect are equal and when we showed that the sum of the angles in a polygon is equal to $(n - 2)\,180°$. Such methods were introduced into general practice by the great Greek mathematician Euclid. They have been accepted ever since to be the only valid and authoritative ones for proving any new theorem. Incidentally, Euclid ended each proof with the phrase "Which proves what we wanted to demonstrate," a phrase abbreviated in Roman times to QED, for *Quod erat demonstrandum* ("Which was to be demonstrated"). This abbreviation became the stamp of authority in mathematics—remaining so to this very day.

When the Four-Color Problem was announced, mathematicians assumed that it could be proved in standard "Euclidean" fashion. Eminent mathematicians such as De Morgan, Arthur Cayley (1821–1895), Arthur Bray Kempe (1849–1922), David Birkhoff (1844–1944), Percy John Heawood (1861–1955), and Philip Franklin (1898–1965) all tried their hand at coming up with the proof. Years passed, but the relevant proof remained elusive.

Then, seemingly out of nowhere, in 1976, two distinguished mathematicians at the University of Illinois, Wolfgang Haken (1928–) and Kenneth Appel (1932–), claimed to have "solved" the Four-Color Problem, not with any of the traditional Euclidean methods of proof but, alas, with a computer program that, they maintained, could "litmus test" any map for the Four-Color conjecture. So far, the program has found no map that requires more than four tints to color distinctively. Haken and Appel wrote a computer program that exhaustedly proved the four-color hypothesis for a critical subset of maps and this, in turn, has been used to imply the hypothesis for all maps. However, many mathematicians are uncomfortable with the Haken-Appel "proof." If accepted as authoritative by one and all, it will have truly altered the face of mathematics. This is perhaps why it is still being debated today.

Mathematical Annotations

The details of the Haken-Appel program and the mathematical principles on which it is based are far too complex to be discussed here. Readers who are interested in them can consult the sources listed in the Further Reading section—particularly the thorough and accessible explanation of the program, as well as the background mathematics needed to understand it, in Robin Wilson's *Four Colors Suffice*. What is pertinent to discuss here is the nature of Euclidean methods and why the Four-Color Problem has had such a profound impact on mathematics.

The Euclidean Method

Proof by deduction has always epitomized mathematical method because of its special capability to demonstrate that some relation or observation holds for the **general case**, an argument or a line of reasoning that refers to a whole category or to every member of a class or a category—to all points, to all angles, to all numbers, and so on. For example, if you measured the angles of different kinds of triangles with a protractor, you would soon start suspecting that the sum is always 180 degrees. But you cannot be sure that the angles in absolutely every triangle will add up to 180 degrees. Proof by deduction allows you to establish this without exception. Here's how. Draw a triangle **ABC** and label its angles a, b, c. Extend its base on both sides, and draw a line parallel to the base, going through the top vertex. Lines parallel to each other are shown with "arrowheads." Note that **ABC** represents any triangle. You could modify it any way you wish (you could make it **obtuse-angled,** right-angled, bigger, smaller, etc.), but the reasoning that follows would still be valid.

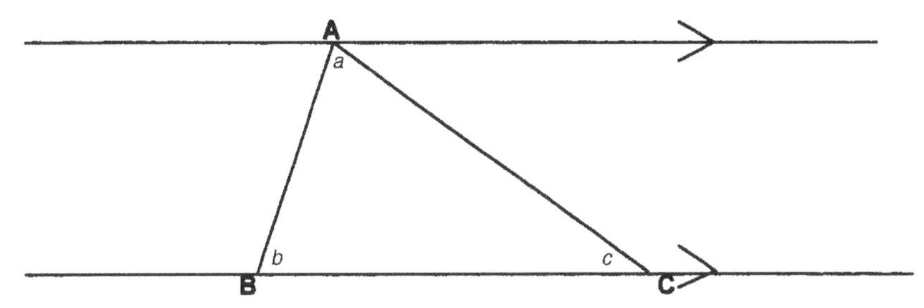

A previously proved theorem of geometry states that when a line crosses two parallel lines, the angles on opposite sides of that line (called a **transversal**) are equal. In our diagram, there are two transversals—**AB** and **AC**. The equal angles that they produce are the one opposite b and the one opposite c. Mark these on the diagram:

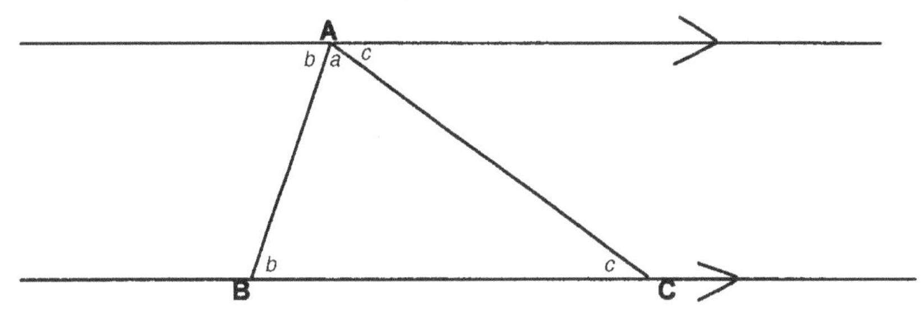

Now, notice that the angle segments a, b, c at vertex **A** are components of a straight angle. Therefore, $a + b + c = 180°$. Notice that the three angles in the triangle are also a, b, and c. We have just established that $a + b + c = 180°$, so we can conclude that the sum of the angles in the triangle is 180 degrees: since a, b, and c can have any value (under 90 degrees, of course), and since the triangle chosen was not given specific dimensions, this demonstration is true under all circumstances. We have thus established beyond a shadow of a doubt that the sum of the angles in *any* triangle will always add up to 180 degrees—QED.

Needless to say, even the Greeks realized that not all theorems in mathematics could be proved by the deductive method. Early on, another method, called **reductio ad absurdum** (literally, "reduced to the absurd"), came to be used in a complementary fashion. This establishes the truth of something by showing that its contradiction is either false or inconsistent.

Actually, this time it was not Euclid, but Zeno of Elea (whom we will meet in chapter 8), who introduced this method into logic. However, Euclid used it ingeniously to prove various theorems, such as the theorem that the set of prime numbers is infinite. As discussed in chapter 3, the integers are divided into primes and composites. The former have no factors that divide into them (other than 1 and themselves); the latter do. The numbers 12, 42, and 169, for instance, are composite. Their factors are

$$12 = 2 \times 2 \times 3$$
$$42 = 7 \times 2 \times 3$$
$$169 = 13 \times 13$$

All composite numbers can be expressed as products of prime factors in this way. Any prime factor of a composite number will thus divide evenly into it.

The first ten primes are: {2, 3, 5, 7, 9, 11, 13, 17, 19, 23}. Even a cursory glance at the set of whole numbers—{0, 1, 2, 3, 4, 5, 6, 7, 8, 9, 10, . . . }—reveals that there are fewer and fewer primes as the numbers increase. Thus, it seems logical to assume that the primes must come to an end at some point. Euclid proved that this is not so.

He started with the assumption that they may indeed come to an end. This means that there is a largest prime. He called it p_n:

$$\text{The complete set of primes} = \{2, 3, 5, 7, \ldots, p_n\}.$$

Euclid then asked, What kind of number would result from multiplying all the primes in the set?

$$\{2 \times 3 \times 5 \times \ldots \times p_n\} = ?$$

As it stands, the number would, of course, be a composite one, made up of all the primes as factors. Any one of them would divide into it. This is a trivial result. So, to make things interesting, Euclid added the number 1 to the product:

$$\{2 \times 3 \times 5 \times \ldots \times p_n\} + 1 = ?$$

That pesky 1 gave Euclid all he needed to shoot down the idea of a largest prime. Why? There are two possibilities for the number that the previous expression represents—it is either prime or composite. Let's assume that it is prime (P):

$$\{2 \times 3 \times 5 \times \ldots \times p_n\} + 1 = P \text{ (a prime number).}$$

P is a number that is, clearly, larger than any in the set {2, 3, 5, . . . p_n}, since it is produced by multiplying "all" of them. It is a "new" prime, and it is much greater than p_n, which turns out, therefore, not to be the largest prime, as originally assumed.

The second possibility is, as mentioned, that the expression yields a composite number (C):

$$\{2 \times 3 \times 5 \times \ldots \times p_n\} + 1 = C \text{ (a composite number).}$$

As a composite number, C is made up of prime factors that will divide into it. But none of the available primes—$\{2, 3, 5, \ldots p_n\}$—will divide into it, because if we take any one of them and divide it into C, that irksome 1 will always be left over as a remainder. So, C must have a prime factor that is not in the set. Again, it is a "new" prime and is greater than any prime in the set. This time, too, it turns out that p_n, the number assumed to be the largest prime, is not the largest one after all. In this way, Euclid proved that the primes never end. He did it by showing that assuming a largest prime produces an "absurdity."

People who tackled the Four-Color Problem at first did so assuming that it could be proved in accordance with such traditional methods of proof. But all efforts to do this turned out to be futile. Haken and Appel's proof is peculiar because it breaks with tradition. The computer program written by Haken and Appel essentially checks to see if a map can be colored by more than four tints. It has convinced many mathematicians that the Four-Color theorem has finally been proved and that it constitutes an important radical innovation in mathematical method. However, as I read the current state of affairs, there continues to be a sense of great unease with accepting Haken and Appel's computer program as "proof," because, technically, it is not really a proof.

TRIANGLES

A right-angled triangle is one in which one of the angles is 90 degrees:

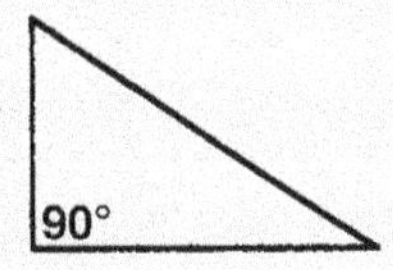

An acute-angled triangle is one in which all three angles are less than 90 degrees:

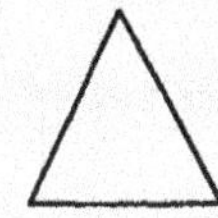

(continued)

An obtuse-angled triangle is one in which one of the angles is greater than 90 degrees:

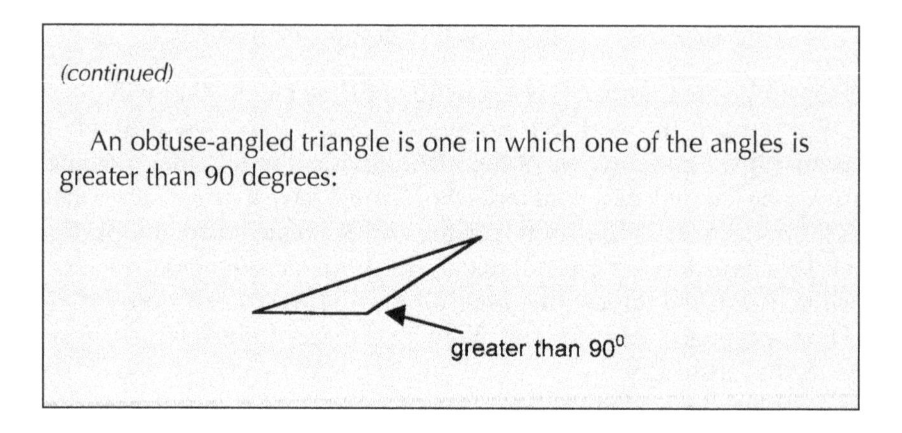

Proof

Among the first to use proof as a way to demonstrate something as being true in general was Pythagoras (chapter 2), who proved that the square of the hypotenuse of any right-angled triangle is equal in area to the squares of the other two sides together, or, in notational form:

$$c^2 = a^2 + b^2.$$

In this equation, c is the length of the hypotenuse and a and b the lengths of the other two sides. For example, if the length of a is 3, then the length of b is 4, and the length of c is 5:

$$5^2 = 3^2 + 4^2$$
$$25 = 9 + 16.$$

The following figure shows this relation in a visual way:

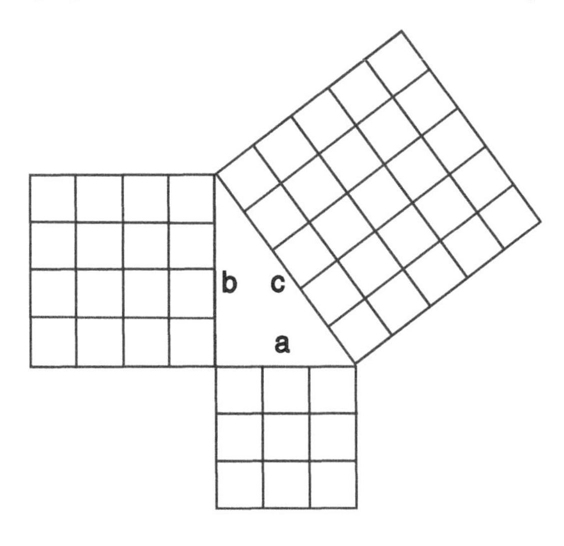

Many cultures throughout the ancient world knew of this relation. History records, however, that Pythagoras was probably the first one to prove that it holds for all right-angled triangles, even though he left no written version of it.

The Pythagorean theorem, as it came to be called, was arguably the critical event that established mathematical method on the technique of proof—a technique that produced insights and results that made it possible to explore other theorems. Sometimes the results even led to unexpected discoveries. The Pythagoreans themselves noticed that their very own theorem, when applied to an isosceles right-triangle with its two sides equal to 1 (the unit length), produced a very strange number for the length of the hypotenuse:

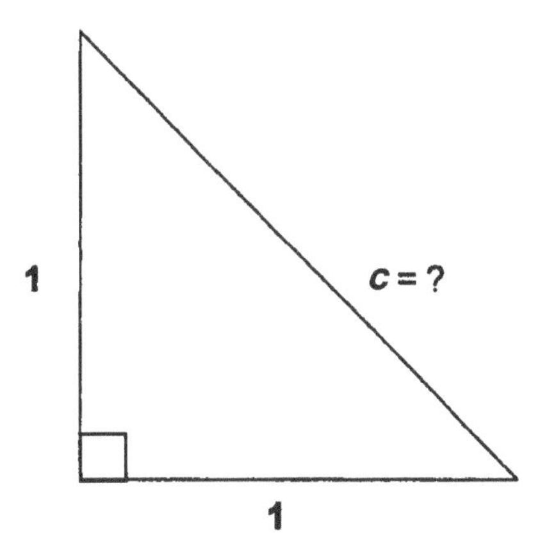

$$c^2 = 1^2 + 1^2 = 1 + 1 = 2.$$

Therefore:

$$c^2 = 2.$$

And thus:

$$c = \sqrt{2}.$$

As it turned out, the number $\sqrt{2}$ cannot be written as a fraction or a *ratio*. It is a repeating decimal (1.4142136 . . .). For this reason, it came to be called *irrational*.

The Pythagoreans were very uncomfortable with their unwitting discovery of irrationals, because of their philosophical beliefs. Euclid, on the other hand, saw irrationals as legitimate numbers. But in order to include them in the growing encyclopedia of mathematical knowledge at the time, he had to prove that they were, in fact, different from rationals. To do so, he used a type of proof that was different from proof by deduction or by reductio ad absurdum. It is called proof by contradiction.

$4n^2$

As discussed in previous chapters, an even number is represented with the formula $2n$. This indicates that if we take any number ($n = 1, 2, 3, \ldots$) and multiply it by 2, we will always get an even number.

The expression $2q^2$ also represents an even number. In this case the n in $2n$ has simply been replaced by q^2.

When we square $2n$, we get $4n^2$: $(2n)(2n) = 4n^2$.

The expression $4n^2$ is itself an even number because $2n$ divides into it. This proves, in effect, that squaring any even number yields a product that is itself an even number.

Simplified Fractions

Some fractions can be simplified (or reduced): for example, the fraction $\frac{5}{10}$ can be simplified to $\frac{1}{2}$ because 5 will divide into both the numerator and the denominator. Similarly, $\frac{4}{6} = \frac{2}{3}$ because 2 will divide into both the numerator and the denominator. Of course, some fractions, such as $\frac{2}{3}$, cannot be simplified, because there is no factor that will divide into both the numerator and the denominator. When a fraction can no longer be simplified, it is said to be in its "lowest" or "reduced" form.

Euclid started by noting that the general form of a rational number is $\frac{p}{q}$. An **integer** is defined as a member of the set of positive whole numbers (1, 2, 3, . . .), negative whole numbers (–1, –2, –3, . . .), and zero. In the case of the integers, the denominator q is always 1—for example, 4 is really $\frac{4}{1}$. Note that q cannot equal 0. Division by zero is not defined. The reason for this will be discussed in chapter 8. Euclid proved simply that $\sqrt{2}$ could not be written in the form $\frac{p}{q}$. He did this by assuming that it could be written in that form and then showing that this would lead to a contradiction.

In order to eliminate the square root sign, Euclid first squared both sides of the relevant equation:

$$\sqrt{2} = \frac{p}{q}$$

$$\left(\sqrt{2}\right)^2 = \frac{p^2}{q^2}.$$

Therefore:

$$2 = \frac{p^2}{q^2}.$$

He then multiplied both sides by q^2, thus eliminating the cumbersome denominator on the right side of the equation:

$$2q^2 = p^2.$$

Now, p^2 is an even number because it equals $2q^2$, which has the form of an even number. It can thus also be concluded that p itself is even. Now, if p is even, we can rewrite it with the general formula for an even number, namely, $p = 2n$, plugging this into the previous equation:

$$2q^2 = p^2.$$

Since $p = 2n$:

$$2q^2 = (2n)^2 = (2n)\,(2n) = 4n^2.$$

Therefore:

$$2q^2 = 4n^2.$$

Now, this equation can be simplified by dividing both sides by 2:

$$q^2 = 2n^2.$$

This equation shows that q^2 is an even number and thus that q itself is an even number and can be written as $2m$ (to distinguish it from $2n$): that is, $q = 2m$. Now, Euclid went right back to his original assumption—namely, that $\sqrt{2}$ was a rational number:

$$\sqrt{2} = \frac{p}{q}.$$

In this equation he substituted what he had just proved, namely, that $p = 2n$ and $q = 2m$:

$$\sqrt{2} = \frac{2n}{2m}.$$

The **fraction** (a numerical expression that indicates the quotient of two quantities, expressed in general form as $\frac{p}{q}$) on the right side can be simplified to $\frac{n}{m}$ (since the common factor 2 can be divided into the numerator and the denominator:

$$\sqrt{2} = \frac{n}{m}.$$

Now, the problem is that we find ourselves back to where we started. We have simply ended up replacing $\frac{p}{q}$ with $\frac{n}{m}$. We could, clearly, continue on in this way, always coming up with a fraction with different numerators and denominators ($\frac{a}{b}, \frac{x}{y}, \ldots$), ad infinitum. But fractions cannot be simplified forever. We have thus reached a contradiction. What caused it? The fact that $\sqrt{2}$ was assumed to have the form $\frac{p}{q}$. It obviously does not. Thus, by contradiction, Euclid proved that $\sqrt{2}$ is not a rational number.

METHODS OF PROOF

Deduction. This shows that something follows necessarily from a set of premises.

Reductio ad Absurdum. This disproves a proposition by showing the absurdity of its inevitable conclusion.

Contradiction. This shows that an original assumption leads to a contradiction and can thus be discarded.

Induction. This shows that something can be established as true if it can be proved for the $(n + 1)^{th}$ case.

To conclude the discussion on proof, it is necessary to mention a fourth type accepted by mathematicians, because it is particularly relevant in any discussion of the Haken-Appel proof of the Four-Color conjecture. It is called proof by induction.

To grasp how this type of proof proceeds, let's return to the formula for summing a sequence that was discussed in chapter 3:

$$\text{Sum}_{(n)} = \frac{n\,(n+1)}{2}.$$

How can this formula be proved? We start by showing that the formula works for the first case, that is, for $n = 1$:

$$\text{Sum}_{(n)} = \frac{n\,(n+1)}{2}$$

$$\text{Sum}_{(1)} = \frac{1 \times (1+1)}{2} = \frac{1 \times \cancel{2}}{\cancel{2}} = 1.$$

This shows that it holds for the first case, because by summing 1, we get 1. The next step is to show that the formula can be applied to a series consisting of one more term after the last. Since the last term is n, the term after it is $(n + 1)$. We label the series $\text{Sum}_{(n+1)}$. The sum of $(n + 1)$ terms can be determined by simply adding the extra term $(n + 1)$ to $\text{Sum}_{(n)}$:

$$\text{Sum}_{(n+1)} = \text{Sum}_{(n)} + (n+1).$$

Since

$$\text{Sum}_{(n)} = \frac{n\,(n+1)}{2}$$

therefore:

$$\text{Sum}_{(n+1)} = \frac{n\,(n+1)}{2} + \frac{(n+1)}{1}.$$

This can be expressed as follows:

$$\text{Sum}_{(n+1)} = \frac{(n+1)\,[(n+1)+1]}{2}.$$

The form of this equation is identical to the form of the one for $\text{Sum}_{(n)}$. Why? Because every appearance of n in $\text{Sum}_{(n)}$ has been replaced by $(n + 1)$, as readers can check for themselves. In other words, we have just shown that the formula is true for $(n + 1)$. Since we can choose n to be as large as we want, we have in effect shown that the formula can be applied to any series. Why? Because we can apply the summation formula not only to the series containing the term after n, but to one containing the term after that, and so on, ad infinitum. Proof by induction can be compared to the "domino effect," whereby a row of dominoes stood on end will fall in succession if the first one is knocked over.

Now, let's go back to the Haken-Appel proof of the Four-Color theorem. Essentially, their program checks to see if any map under consideration can be colored with more than four colors. Their "proof" is thus not a real proof in the traditional sense of the word. It is a set of computer instructions. How can we be sure that the instructions devised by Haken and Appel will apply to all maps ad infinitum? In traditional mathematical terms, how can we be sure that their "proof" entails the domino effect, in the same way that proof by induction does?

In my view, the Four-Color Problem remains a true puzzle for many (maybe most) mathematicians. Its solution in traditional mathematical terms is probably still "out there" and will hinge on some "Aha!" insight that has not as yet come about. Originally, the Pythagorean theorem was a conjecture—something that could be shown to hold in specific cases without ever finding an exception. It became a true "theorem" the instant that it could be shown to hold for the general case. The Four-Color Problem will continue probably to bother many mathematicians until some simple proof of the general case is found. However, as the great American philosopher

Charles S. Peirce (1839–1914), who came under its spell, so aptly put it in a lecture he delivered at Harvard in the 1860s, the problem is so infuriating precisely because it appears to be so simple to prove, and yet no mathematician has ever found a proof for it with the traditional methods of logic and mathematics.

Reflections

The Four-Color Problem has opened up a veritable "can of worms" in mathematical circles. If the Haken-Appel proof is accepted as is, then it constitutes a true innovation in mathematical method. However, since the proof cannot be examined in the same way that proofs have been examined traditionally, many mathematicians feel very uneasy about it. Certainly, it fails to conform to the Greek ideal of certain, absolute proof. The most we can say is that it is "probably" true, like the theories of physicists.

Maybe some day the insight leading to the elusive simple proof that Peirce referred to may still emerge. As Haken and Appel themselves have admitted (from Appel and Haken, 2002, in Further Reading): "One can never rule out the chance that a short proof of the Four-Color theorem might some day be found, perhaps by the proverbial bright high-school student." My sense is that many mathematicians are still waiting for that student to step forward!

Explorations

Coloring Problems

42. What is the least number of colors necessary to completely fill in each of the following maps? Any two neighboring regions must be filled in with different colors.

A.

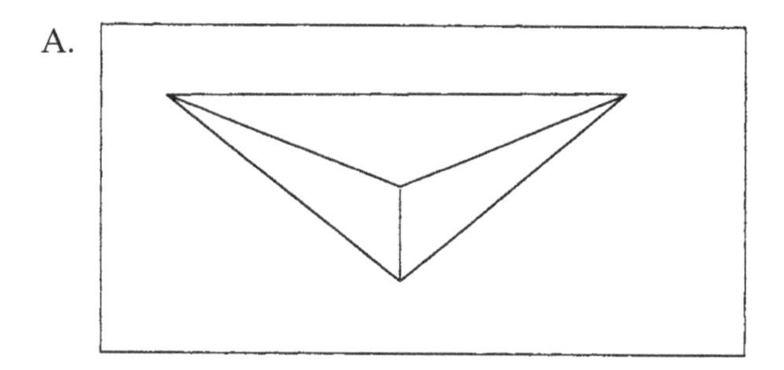

B.

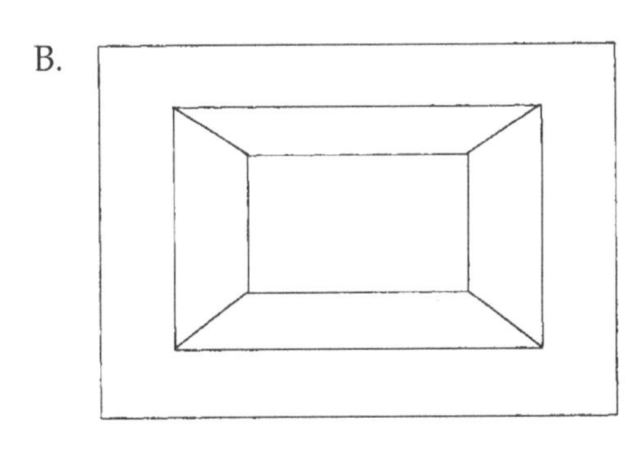

C.

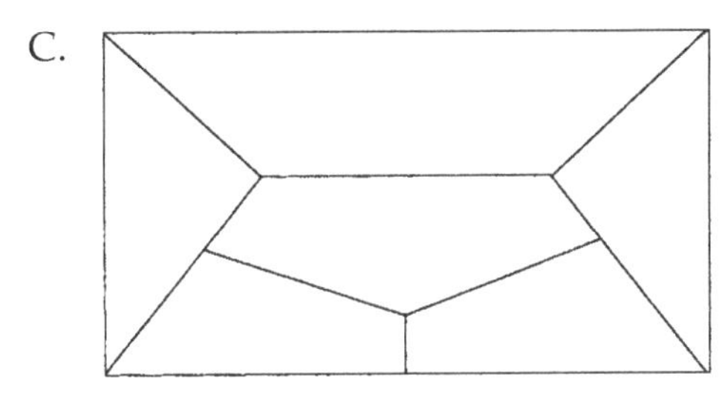

D.

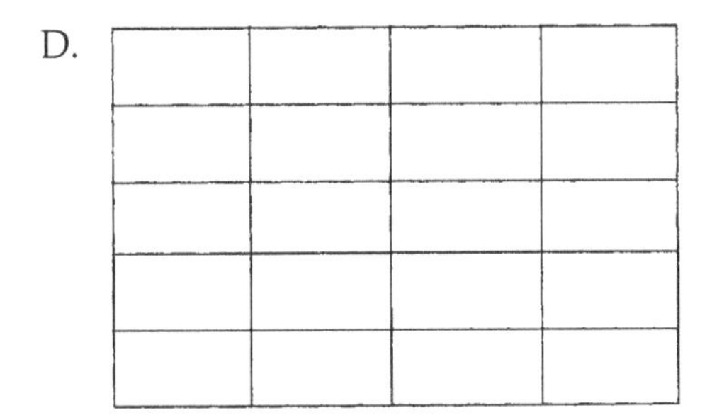

43. What is the least number of colors necessary to color a Möbius strip and a Klein bottle (chapter 4)?

Proof

44. Prove by the method of contradiction that if one of the angles is greater than 90 degrees in a triangle, then the other two must each be less than 90 degrees.

45. Prove that a cube is a network to which the formula for three-dimensional graphs, $v - e + f = 2$, can be applied (chapter 4).

46. Prove that the exterior angle of a triangle, angle x, in the following figure, is equal to the sum of the internal opposite angles, y and z.

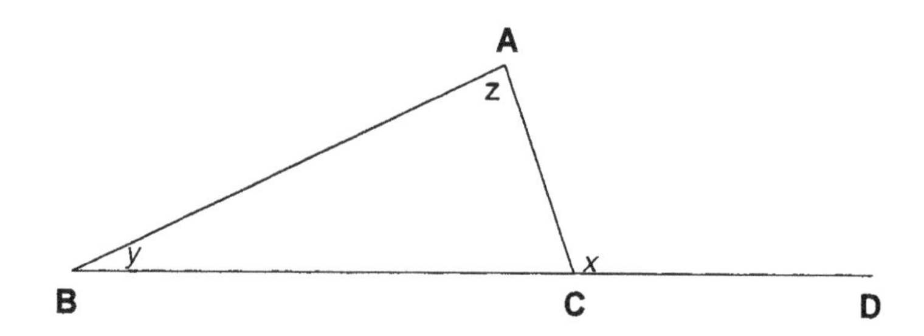

47. How many colors are needed to color the following figure? Can this be proved to be the case, without actually having to color it?

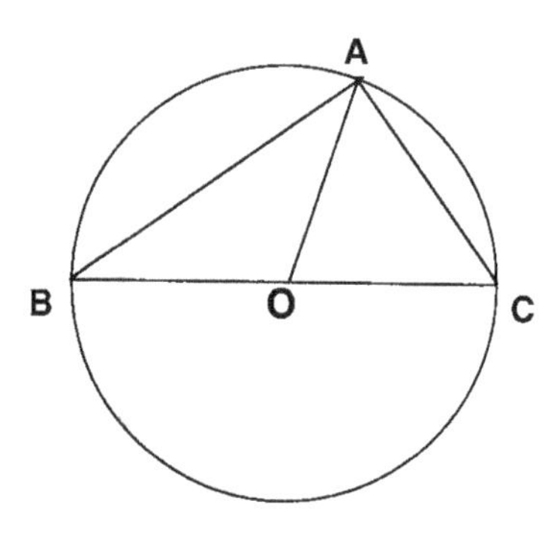

Further Reading

Appel, Kenneth, and Wolfgang Haken. "The Four-Color Problem." In Dale Jacquette (ed.), *Philosophy of Mathematics*, 193–208. Malden, Mass.: Blackwell, 2002.

———. "The Four-Color Proof Suffices." *The Mathematical Intelligencer* 8 (1986): 10–20.

Barnette, David. *Map-Coloring Polyhedra and the Four-Color Problem*. Washington, D.C.: Mathematical Association of America, 1983.

Benson, D. C. *The Moment of Proof: Mathematical Epiphanies*. Oxford: Oxford University Press, 1999.

Casti, John L. *Mathematical Mountaintops: The Five Most Famous Problems of All Time*. Oxford: Oxford University Press, 2001.

Doxiadis, A. *Uncle Petros and Goldbach's Conjecture.* London: Faber and Faber, 2000.

Haken, Wolfgang, and Kenneth Appel. "The Solution of the Four-Color-Map Problem." *Scientific American* 237 (1977): 108–21.

Jacquette, Dale (ed.). *Philosophy of Mathematics.* Malden, Mass.: Blackwell, 2002.

Tymoczko, Thomas. "The Four-Color Problem and Its Philosophical Significance." *Journal of Philosophy* 24 (1978): 57–83.

Wilson, Robin. *Four Colors Suffice: How the Map Problem Was Solved.* Princeton, N.J.: Princeton University Press, 2002.

Lucas's Towers of Hanoi Puzzle

There is no more steely barb than that of the Infinite.

CHARLES BAUDELAIRE (1821–1867)

GAMES PLAYED ACCORDING TO RULES, and involving equipment such as boards, counters, sticks, stones, coins, and the like, have intrigued and entertained people of all ages from time immemorial. Speculation abounds as to why they were invented in the first place. But the question of their purpose has never been satisfactorily answered.

Mathematicians, too, have always been intrigued by games, because many of these can be constructed on the basis of mathematical principles

FRANÇOIS ANATOLE LUCAS (1842–1891)

Lucas was educated at the École Normale in Amiens. During the Franco-Prussian War (1870–1871), he served as an artillery officer. After the war, he was hired to teach mathematics at the Lycée Saint Louis in Paris. Lucas later taught at the Lycée Charlemagne, also in Paris. He died as the result of a freak accident at a banquet, when a plate was dropped and a piece flew up and cut his cheek, lethally infecting it.

Lucas is best known for his work in number theory. As we saw in chapter 3, he studied the Fibonacci sequence and the implications it had for mathematics.

and thus used as "test devices" for modeling those very principles. One of the most famous and fascinating of all such mathematically designed games is known as the Towers of Hanoi Puzzle. It was invented as a toy for children in 1883 by the French mathematician François Edouard Anatole Lucas, whom we encountered in chapter 3, although the idea pattern it embodies goes back considerably in time and is found in cultures throughout the world. The puzzle is, in effect, a "toy model" of the concept of geometric series. Given its simplicity and ingenuity, and the fact that it continues to intrigue us to this day, it qualifies as one of the ten greatest puzzles of all time. Even now, a simplified version of the puzzle can be found in toy stores everywhere:

Before getting to the puzzle itself, it is useful to revisit briefly the concept of series. As discussed in chapter 3, a series in mathematics is defined as an ordered succession of numbers:

1. $\{2, 4, 6, 8, 10, 12, 14, 16, \ldots\}$
2. $\{2, 4, 8, 16, 32, 64, 128, \ldots\}$

In series 1, known as an arithmetical series, each term is greater by 2 than the one just before it. Let's now generalize the construction of such a series. If we use a to represent the initial term of the series and d to stand for the constant difference between two successive terms, then the general form of an arithmetic series can be set up as follows:

$$\{a, a + d, a + 2d, a + 3d, \ldots, a + (n - 1)d\}.$$

In series 1, $a = 2$ and $d = 2$. Plugging these values into the successive terms of the general form will yield the actual terms of our series:

First Term	Second Term	Third Term	Fourth Term	...
↓	↓	↓	↓	
a	$a + d$	$a + 2d$	$a + 3d$	...
↓	↓	↓	↓	
2	$2 + 2 = 4$	$2 + (2 \times 2) = 6$	$2 + (3 \times 2) = 8$	...

The expression $a + (n - 1)d$ indicates that any term in an arithmetical series is constructed with the initial term a plus the constant difference $(n - 1)d$, where n is the positional number (1st, 2nd, 3rd, . . .) of a term in the series:

TABLE 6-1: THE GENERAL TERM OF AN ARITHMETICAL SERIES

Term	Form	Method of Construction
1st	a	$a + (1 - 1)d = a + (0)d = a$
2nd	$a + d$	$a + (2 - 1)d = a + (1)d = a + d$
3rd	$a + 2d$	$a + (3 - 1)d = a + (2)d = a + 2d$
4th	$a + 3d$	$a + (4 - 1)d = a + (3)d = a + 3d$
. . .	. . .	. . .
nth	$a + (n - 1)d$	$a + (n - 1)d$

In series 2, known as a geometric series, each term is formed by multiplying the one just before it by 2, called the common ratio. Let's generalize the structure of this type of series as well. If we use a again to represent the initial term of the series and r to stand for the common ratio, then the general form of a geometric series can be set up as follows:

$$\{a, ar, ar^2, ar^3, ar^4, \ldots, ar^{n-1}\}.$$

In series 2, $a = 2$ and $r = 2$. Plugging these values into the successive terms of the general form will yield the actual terms in series 2:

First Term	Second Term	Third Term	Fourth Term	...
↓	↓	↓	↓	...
a	ar	ar^2	ar^3	...
↓	↓	↓	↓	
2	$2 \times 2^1 = 4$	$2 \times 2^2 = 8$	$2 \times 2^3 = 16$	...

The expression ar^{n-1} represents the general term of a geometric series. It indicates that any term in the series is constructed with the initial term a multiplied by r^{n-1}, where n is the positional number (1st, 2nd, 3rd, . . .) of a term in the series:

TABLE 6-2: THE GENERAL TERM OF A GEOMETRIC SERIES

Term	Form	Method of Construction
1st	a	$ar^{1-1} = ar^0 = a$
2nd	ar	$ar^{2-1} = ar^1 = ar$
3rd	ar^2	$ar^{3-1} = ar^2$
4th	ar^3	$ar^{4-1} = ar^3$
. . .	. . .	. . .
nth	ar^{n-1}	ar^{n-1}

The basic ideas in the foregoing discussion will come in handy in grasping the nature of Lucas's puzzle.

EXPONENTS

Multiplication

Multiplying the same digits with exponents is equivalent to adding their exponents, as in these examples:

$$3^4 \times 3^5 = 3^{4+5} = 3^9$$
$$7^{12} \times 7^{20} = 7^{12+20} = 7^{32}$$

Why? Take the first example:

$$\begin{array}{ccc} 3^4 & \times & 3^5 \\ \downarrow & & \downarrow \\ (3 \times 3 \times 3 \times 3) & \times & (3 \times 3 \times 3 \times 3 \times 3). \end{array}$$

The exponents tell us, in effect, how many factors are involved in the multiplication. Counting the number of factors, you will get 9—which is the same number as adding up the number of exponents ($4 + 5 = 9$).

So, in general:

$$a^n \times a^m = a^{n+m}.$$

Division

Dividing the same digits with exponents is equivalent to subtracting their exponents. For example:

$$3^5 \div 3^3 = 3^{5-3} = 3^2$$
$$7^{15} \div 7^5 = 7^{15-5} = 7^{10}$$

(continued)

Why? Take the first example. Again, the exponents tell us how many factors are involved in the division:

$$\frac{3^5 = (3 \times 3 \times 3) \times 3 \times 3}{3^3 = (3 \times 3 \times 3)}.$$

Canceling out the factors in parentheses, we get

$$\frac{\cancel{(3 \times 3 \times 3)} \times 3 \times 3}{(\cancel{3 \times 3 \times 3})}.$$

This leaves 3×3, or 3^2—which is the same number as subtracting the two exponents ($5 - 3 = 2$).

So, in general:

$$a^n \div a^m = a^{n-m}.$$

$n^0 = 1$

Any number raised to the power of 0 is equal to 1.

Why? Take two identical digits that have the same exponent and divide them: $3^5 \div 3^5$.

We know from the previous examples that $3^5 \div 3^5 = 3^{5-5} = 3^0$.

But the result of dividing 3^5 by 3^5 is 1:

$$\frac{3^5 = \cancel{3 \times 3 \times 3 \times 3 \times 3}}{3^5 = \cancel{3 \times 3 \times 3 \times 3 \times 3}}.$$

So, $3^5 \div 3^5 = 1$.

Since $3^5 \div 3^5$ equals 3^0, we have thus proved that $3^0 = 1$.

In general:

$$n^0 = 1.$$

The Puzzle

As mentioned, the Towers of Hanoi Puzzle appeared in 1883. Lucas probably got the idea for it from a similar problem included in the 1550 edition of *De Subtililate*, by the Italian mathematician Girolamo Cardano (1501–1576):

> A monastery in Hanoi has a golden board with three wooden pegs on it. The first of the pegs holds sixty-four gold disks in descending order of size—the largest at the bottom, the smallest at the top. The

> monks have orders from God to move all the disks to the third peg while keeping them in descending order, one at a time. A larger disk must never sit on a smaller one. All three pegs can be used. When the monks move the last disk, the world will end. Why?

The world would end because it would take the monks $2^{64} - 1$ moves to accomplish the task of moving the disks as stipulated. Even at one move per second (and no mistakes), the task would require 5.82×10^{11}, or 582,000,000,000, years to accomplish! Before explaining this, it is useful to briefly introduce the concept of exponents.

Consider multiplying 3 by itself fourteen times:

$$3 \times 3 \times 3 \times 3 \times 3 \times 3 \times 3 \times 3 \times 3 \times 3 \times 3 \times 3 \times 3 \times 3.$$

This layout of the multiplication is clearly cumbersome and thus inefficient to work with. To make multiplication more "economical," mathematicians have come up with the concept of **exponent**. The previous multiplication can be represented more efficiently as 3^{14}, where 3 is called the **base** or **root**, and the superscript "^{14}" the exponent or **power**. The exponent tells us the number of times that the base is to be multiplied by itself. In general, n^m indicates that n is to be multiplied by itself m times. In 2^8, for example, $n = 2$ and $m = 8$:

$$2^8 = 2 \times 2 \times 2 \times 2 \times 2 \times 2 \times 2 \times 2.$$

Note that a number to the zero power is equal to 1, no matter what it is:

$2^0 = 1$, $4^0 = 1$, $13^0 = 1$, and so on.
(For proof of this, see the sidebar on page 109.)

Exponents are particularly useful for representing terms in a geometric series. For example, in a series consisting of terms that are successive powers of 2, the last term would be 2^n:

$$\{2^0, 2^1, 2^2, 2^3, 2^4, 2^5, \ldots, 2^n\}.$$

The term just before 2^n would, of course, be 2^{n-1}, and the one before that, 2^{n-2}:

$$\{2^0, 2^1, 2^2, 2^3, 2^4, 2^5, \ldots, 2^{n-2}, 2^{n-1}, 2^n\}.$$

Now, we are ready to tackle the Towers of Hanoi Puzzle. We start with the simplest version of the puzzle—one in which two disks are to be moved from the first peg to the third one, keeping them in descending order, that is, without a larger disk ever sitting on a smaller one. To keep track of the moves, it is useful to number the disks:

1 = the smaller disk

2 = the larger disk

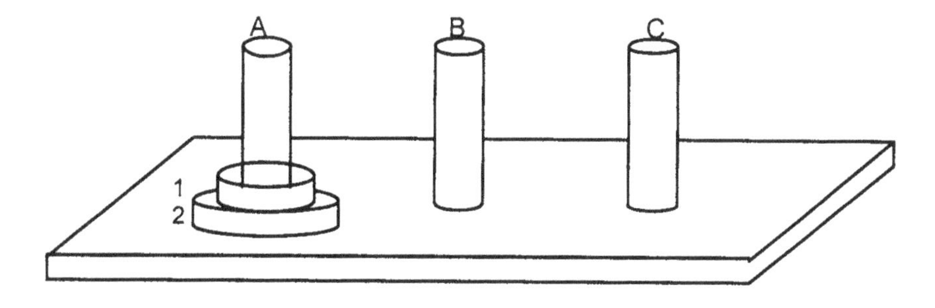

We start by moving disk 1 to peg B:

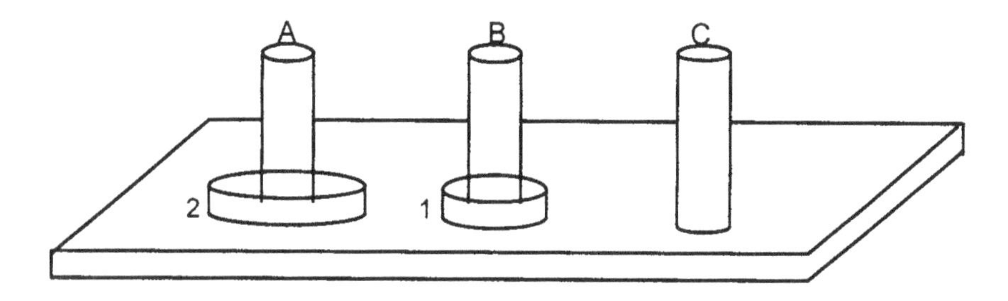

Now, we can move disk 2 on A over to C:

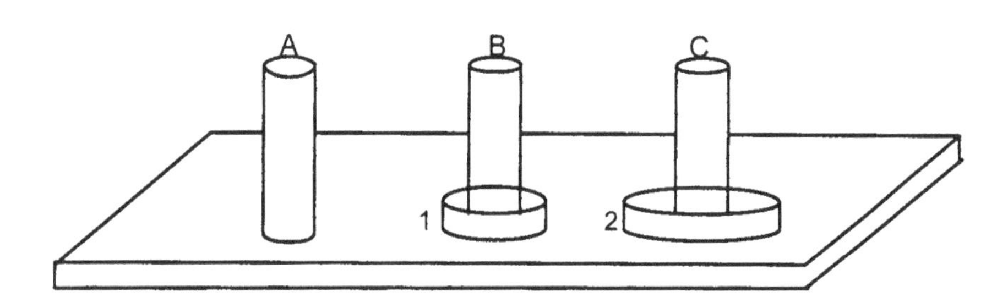

Finally, we move disk 1 to peg C on top of 2, at which point the two disks have been transferred to C, with the smaller one on top, as required:

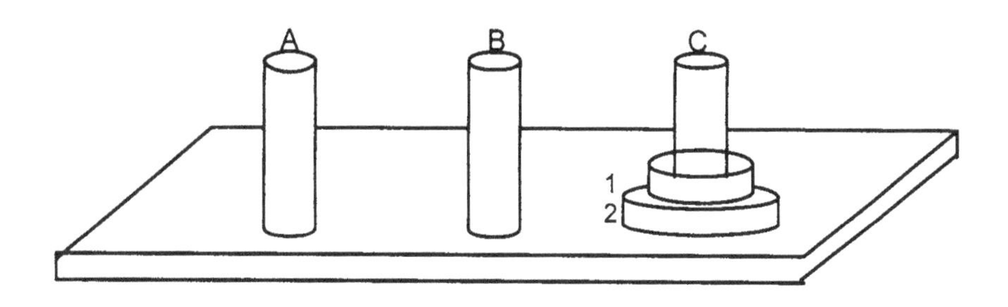

It took three moves to accomplish the task. Note that this result can be represented as $2^2 - 1$, because $2^2 - 1 = 4 - 1 = 3$. Note as well that the exponent "2" in $2^2 - 1$ stands for the number of disks in the game.

Now, consider a three-disk version of the game. We start by numbering the disks 1, 2, and 3:

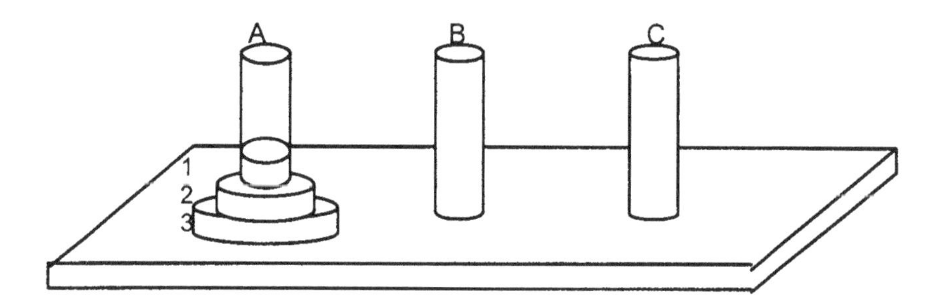

The moves (given without diagrams) are as follows. Readers who may have difficulty envisioning each move should make a physical model of the game, reproducing the moves on the model. You can also use commercially available toy versions of the game:

1. Move disk 1 from A to C.
2. Move disk 2 from A to B.
3. Move disk 1 from C to B on top of 2.
4. Move disk 3 from A to C (which is now empty).
5. Move disk 1 from B to A (which is now empty).
6. Move disk 2 from B to C on top of 3.
7. Move disk 1 from A to C on top of 2, which is itself on top of 3.

This time, it took seven moves to accomplish the task. Note that this result can be represented as $2^3 - 1$, because $2^3 - 1 = 8 - 1 = 7$. As in the previous two-disk version, note as well that the exponent "3" stands for the number of disks in the game.

It is already obvious that a general pattern is probably involved. If we were to play the Towers of Hanoi game with four, five, and higher numbers of disks, we would in fact find that the number of moves increases according to the general formula $2^n - 1$. In that formula, n represents the number of disks. In Lucas's puzzle, the number of disks is $n = 64$, so the number of moves needed to accomplish the task of transferring the disks from the first to the third peg would be $2^n - 1 = 2^{64} - 1$ which, as mentioned earlier, is an astronomical figure.

Here is a summary of the sixty-four versions of the game—that is, of versions of the game played with one, two, three, and so on, up to sixty-four disks:

TABLE 6-3: TOWERS OF HANOI—SIXTY-FOUR VERSIONS

Disks	**Number of Moves Required $2^n - 1$ (n = number of disks)**						
1	$2^n - 1$	=	$2^1 - 1$	=	(2 – 1)	=	1
2	$2^n - 1$	=	$2^2 - 1$	=	(4 – 1)	=	3
3	$2^n - 1$	=	$2^3 - 1$	=	(8 – 1)	=	7
4	$2^n - 1$	=	$2^4 - 1$	=	(16 – 1)	=	15
5	$2^n - 1$	=	$2^5 - 1$	=	(32 – 1)	=	31
6	$2^n - 1$	=	$2^6 - 1$	=	(64 – 1)	=	63
7	$2^n - 1$	=	$2^7 - 1$	=	(128 – 1)	=	127
. . .	. . .						
64	$2^n - 1$	=	$2^{64} - 1$	=	a very large number!		

To put it in strict mathematical terms, each of the moves required in successive versions of the game turns out to be a successive term in a geometric series with the final term ($2^n - 1$):

$$\{(2^1 - 1), (2^2 - 1), (2^3 - 1), (2^4 - 1), \ldots, (2^n - 1)\} = \{1, 3, 7, 15, 31, \ldots\}.$$

As can be seen, Lucas's puzzle is a simple, albeit "dramatic," illustration of the enormity of "exponential" growth.

Mathematical Annotations

The notion of exponential growth has captured the fancy of many puzzlists throughout history. In 1256, the Arab mathematician Ibn Kallikan cleverly used a chessboard to illustrate it. His puzzle is paraphrased as follows:

> How many grains of wheat are needed on the last square of a sixty-four-square chessboard if one grain is to be put on the first square of the board, two on the second, four on the third, eight on the fourth, and so on in this fashion?

Like Lucas's puzzle, this one also produces a geometric series: $\{2^0, 2^1, 2^2, 2^3, 2^4, \ldots, 2^{63}\}$. Each term stands for the number of grains on each successive square of the chessboard:

On the first square: 1 grain = 2^0 grains

On the second square: 2 grains = 2^1 grains

On the third square: 4 grains = 2^2 grains

On the fourth square: 8 grains = 2^3 grains

On the fifth square: 16 grains = 2^4 grains

. . .

On the sixty-fourth square: = 2^{63} grains

Here is what the chessboard would look like with grains up to the eighth square ($2^7 = 128$):

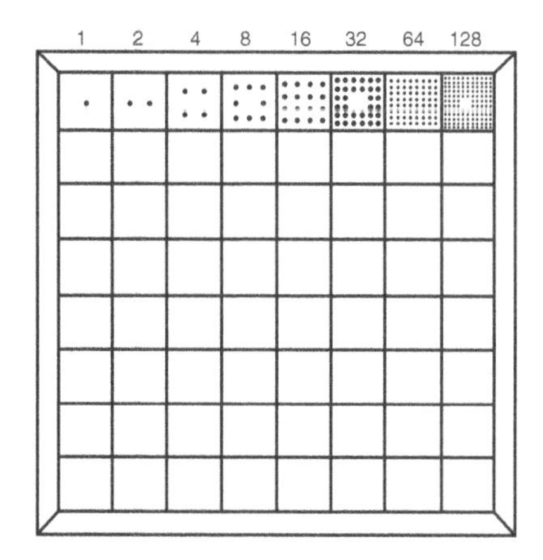

If we use n to represent the number of a term in the series, then the last (sixty-fourth) term is (2^{n-1}). This indicates that the power of the term representing the square on the chessboard is one less than the number of the square. The value of 2^{63} is so large that it boggles the mind to think of what kind of chessboard could hold so many grains, not to mention where so much wheat could be found. The sixty-fourth square would contain about 1.84×10^{19} grains. This figure amounts to some 3×10^{13} bushels, which is several times the world's annual crop of wheat!

Perfect Numbers and Mersenne Primes

Ibn Kallikan's puzzle conceals some truly intriguing patterns. For example, if a second chessboard is placed next to the first, then the pile on the last square (128th square) of the second board contains 2^{127} grains. If we subtract the number 1 from this, $(2^{127} - 1)$ we get the following result: 170,141,183,460,231,731,687,303,715,884,105,727. Incredibly, this is a prime number!

You may have noticed that subtracting one from the number of grains on any of the chessboard squares is equivalent to representing the square with the Towers of Hanoi formula: $(2^n - 1)$. This very formula has an interesting history behind it. It was used, for instance, by Euclid to generate so-called perfect numbers.

A **perfect number** is defined as an integer that equals the sum of all its divisors except itself. The smallest perfect number is 6, which has three divisors, 1, 2, and 3 ($6 = 1 \times 2 \times 3$), and these add up to 6 ($= 1 + 2 + 3$). The next perfect number is 28, whose divisors (1, 2, 4, 7, 14) add up to 28 ($= 1 + 2 + 4 + 7 + 14$). Perfect numbers have fascinated people throughout the ages. In his *City of God*, St. Augustine (354–430) argued that God took six days to create the world, resting on the seventh, because 6, as a perfect number, symbolized the perfection of creation. Incidentally, the next three perfect numbers after 6 are: 28 (as we have just seen), 496, and 8,128. It took nearly 1,400 years after their discovery in ancient Greece before the fifth one was discovered. It is 33,550,336. To the best of my knowledge, only seventeen perfect numbers have so far been discovered. The last one has 1,373 digits and would fill this page if written out.

Euclid claimed that the formula $[2^{n-1}(2^n - 1)]$ would generate all the perfect numbers. But, as it turns out, it generates only even perfect numbers when the expression $(2^n - 1)$ in it is a prime number—a fact proved by Leonhard Euler two millennia later. For example, if $n = 2$, then $(2^n - 1) = (2^2 - 1) = (4 - 1) = 3$. Since this is a prime number, we can now use Euclid's formula to generate a perfect number:

$$[2^{n-1}(2^n - 1)] = [2^{2-1}(2^2 - 1)] = [2^1 (4 - 1)] = (2)(3) = 6.$$

No odd perfect numbers have ever been found. They probably do not exist.

Now, let's take a closer look at Ibn Kallikan's chessboard through the template of Euclid's formula. Recall that the successive numbers of grains on each square can be represented with the formula 2^n, starting with $n = 0$:

2^0	2^1	2^2	2^3	2^4	2^5	2^6	2^7
2^8	2^9	2^{10}	2^{11}	2^{12}	2^{13}	2^{14}	2^{15}
2^{16}	2^{17}	2^{18}	2^{19}	2^{20}	2^{21}	2^{22}	2^{23}
2^{24}	2^{25}	2^{26}	2^{27}	2^{28}	2^{29}	2^{30}	2^{31}
2^{32}	2^{33}	2^{34}	2^{35}	2^{36}	2^{37}	2^{38}	2^{39}
2^{40}	2^{41}	2^{42}	2^{43}	2^{44}	2^{45}	2^{46}	2^{47}
2^{48}	2^{49}	2^{50}	2^{51}	2^{52}	2^{53}	2^{54}	2^{55}
2^{56}	2^{57}	2^{58}	2^{59}	2^{60}	2^{61}	2^{62}	2^{63}

If we then take away one grain from each square, the result would be, as mentioned, $(2^n - 1)$. As it turns out, this formula can be used to test the primality of each square. For example, the fourth square has 2^3 or 8 grains on it. If we take one away from it, $(2^3 - 1)$, we get 7, which is a prime number. Primes derived in this way are called **Mersenne primes**, also known as **Mersenne numbers,** after the French mathematician Marin Mersenne (1588–1648), who used the formula $(2^n - 1)$ as a test of primality. Applied to Ibn Kallikan's chessboard, the formula produces prime numbers on the squares shaded above:

Square	Value of Square	Mersenne Value of the Square	Prime Value
3rd	$2^2 = 4$	$(2^n - 1) = 2^2 - 1$	3
4th	$2^3 = 8$	$(2^n - 1) = 2^3 - 1$	7
6th	$2^5 = 32$	$(2^n - 1) = 2^5 - 1$	31
8th	$2^7 = 128$	$(2^n - 1) = 2^7 - 1$	127
Etc.			

The Mersenne test of primality has been used to determine large primes. In 1978, for instance, two high school students in California, Laura Nickel and Curt Landon Noll, found that $(2^{21,701} - 1)$ was prime, using computer techniques. It was the twenty-fifth Mersenne prime to have been discovered. It has 6,533 digits. In 1996, a loose international Internet alliance of prime number aficionados, founded by the computer programmer George Woltman in Florida, known as GIMPS (the Great Internet Mersenne Prime Search Project), determined that $(2^{3,021,377} - 1)$ was prime. It was the thirty-seventh Mersenne prime discovered. It has 909,526 digits. In 1999, the group discovered another Mersenne prime, $(2^{6,972,593} - 1)$, a number that has 2,098,960 digits. Interested readers should note the Internet address of GIMPS: www.mersenne.org.

Infinity

The search for larger and larger primes raises the question of mathematical *infinity*. The ancient Greeks certainly knew about the value of studying infinity, as we shall see in chapter 8. But it was the German mathematician Georg Cantor who made the study a branch of mathematics at the threshold of the twentieth century.

The great Italian scientist Galileo Galilei (1564–1642) suspected that mathematical infinity posed a serious challenge to common sense. In his 1632 *Dialogue Concerning the Two Chief World Systems*, he noted that the set of square integers can be compared, one by one, with all the whole numbers (positive integers), leading to the preposterous possibility that there may be as many square integers as there are numbers (even though the squares are themselves only a part of the set of integers).

GEORG CANTOR (1845–1918)

Cantor was born in Saint Petersburg (Russia) of German parents. His early work with series led to his development of **set theory** (the study of the properties of sets), upon which modern mathematical analysis is based. His work on infinite series shook the foundations of mathematics, and they are still trembling somewhat.

How can this be, in view of the fact that there are numbers that are not squares, as the following comparison of the two sets seems to show?

Integers:	1	2	3	4	5	6	7	8	9	10	11	12	...
	↕	↕	↕	↕	↕	↕	↕	↕	↕	↕	↕	↕	
Squares:	1	—	—	4	—	—	—	—	9	—	—	—	...

As one would expect, this comparison makes it obvious that there are many more "blanks" in the bottom set (the set of square integers), given that it is a subset of the top set (the set of whole numbers). Common sense would lead us to conclude that the set of whole numbers has thus many more members in it than the set of square numbers does. But it does not. In 1872, Cantor revisited Galileo's insight and showed that it was accurate after all—the two sets have the same number of elements. This can be shown simply by eliminating the blanks in the previous layout and putting the square integers in a direct one-to-one correspondence with the complete set of whole numbers. The result shows that there are no "leftovers," no matter how far we extend the comparison. Every whole number can be matched with exactly one square number, and vice versa:

Integers:	1	2	3	4	5	6	7	8	9	10	11	12	...
	↕	↕	↕	↕	↕	↕	↕	↕	↕	↕	↕	↕	
Squares:	1	4	9	16	25	36	49	64	81	100	121	144	...
	↕	↕	↕	↕	↕	↕	↕	↕	↕	↕	↕	↕	
	1^2	2^2	3^2	4^2	5^2	6^2	7^2	8^2	9^2	10^2	11^2	12^2	...

Now, what is even more bizarre is the fact that the same one-to-one correspondence can be set up between the whole numbers and numbers raised to any power:

Integers:	1	2	3	4	5	6	7	8	9	10	11	12	...
	↕	↕	↕	↕	↕	↕	↕	↕	↕	↕	↕	↕	
Powers:	1^n	2^n	3^n	4^n	5^n	6^n	7^n	8^n	9^n	10^n	11^n	12^n	...

This simple but brilliant comparison technique puts a "fly in the logical ointment of common sense," so to speak. Cantor's argument was, in fact, earth-shattering in mathematical circles when he first made it public. Its aftershocks are still being felt today.

The study of mathematical infinity is filled with paradoxes. For example, a one-to-one correspondence can be set up between the set of "counting numbers" and any of its subsets. Two cases in point are the even and the odd numbers:

Integers:	1	2	3	4	5	6	7	8	9	10	11	12	...
	↕	↕	↕	↕	↕	↕	↕	↕	↕	↕	↕	↕	
Even Integers:	2	4	6	8	10	12	14	16	18	20	22	24	...

Integers:	1	2	3	4	5	6	7	8	9	10	11	12	...
	↕	↕	↕	↕	↕	↕	↕	↕	↕	↕	↕	↕	
Odd Integers:	1	3	5	7	9	11	13	15	17	19	21	23	...

Because the counting numbers are also called **cardinal numbers** (positive integers), any set of numbers that can be put in a one-to-one correspondence with them is said to have the same **cardinality**. Cantor used this notion to investigate all kinds of sets. Consider the set of rational numbers. As we saw in the previous chapter, these are numbers that can be written in the form $\frac{p}{q}$ where p and q are integers (and $q \neq 0$). Thus, for instance, $\frac{2}{3}$, $\frac{-5}{8}$, 5, and $\frac{4}{7}$ are rational numbers. The cardinal numbers are, themselves, a subset of the rationals—every integer p can be written in the form $\frac{p}{1}$. Terminating decimal numbers are also rational, because a number such as 3.579 can be written in $\frac{p}{q}$ form as 3,579/1,000. Finally, all repeating decimal numbers are rational, although the proof of this is beyond the scope of the present discussion. For example, 0.3333333... can be written as $\frac{1}{3}$.

Amazingly, Cantor demonstrated that the rationals also have the same cardinality as the counting numbers. His method of proof is, again, unexpectedly elegant and simple. First, he arranged the set of all rational numbers as shown in the following array:

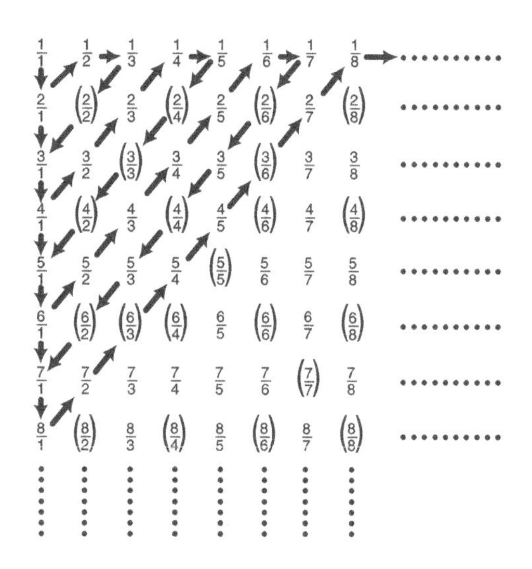

In each row, the successive denominators (q) are the integers {1, 2, 3, 4, 5, 6, . . . }. The numerator (p) of all the numbers in the first row is 1; of all those in the second row, 2; of all those in the third row, 3; and so on. In this way, all numbers of the form $\frac{p}{q}$ are covered in the previous array. Cantor enclosed in parentheses every fraction in which the numerator and the denominator have a common factor. If these fractions are deleted, then every rational number appears once and only once in the array. Now, Cantor set up a one-to-one correspondence between the integers and the numbers in the array as follows: he let the cardinal number 1 correspond to $\frac{1}{1}$ at the top left-hand corner of the array; 2 to the number below $\frac{2}{1}$; following the arrow, he let 3 correspond to $\frac{1}{2}$; following the arrow, he let 4 correspond to $\frac{1}{3}$; and so on, ad infinitum. The path indicated by the arrows therefore allows us to set up a one-to-one correspondence between the cardinal numbers and all the rational numbers (eliminating the numbers in parentheses):

Integers:	1	2	3	4	5	6	7	8	9	10	11	12	13	. . .
	$\updownarrow$	$\updownarrow$	$\updownarrow$	$\updownarrow$	$\updownarrow$	$\updownarrow$	$\updownarrow$	$\updownarrow$	$\updownarrow$	$\updownarrow$	$\updownarrow$	$\updownarrow$	$\updownarrow$	
Array Numbers:	$\frac{1}{1}$	$\frac{2}{1}$	$\frac{1}{2}$	$\frac{1}{3}$	$\frac{3}{1}$	$\frac{4}{1}$	$\frac{3}{2}$	$\frac{2}{3}$	$\frac{1}{4}$	$\frac{1}{5}$	$\frac{5}{1}$	$\frac{6}{1}$	$\frac{5}{2}$	. . .

Conclusion? There are as many rational numbers as there are whole numbers! One cannot help but be impressed by the elegant and simple way in which Cantor constructed this mind-boggling demonstration. In effect, once the simplicity inherent in the principles of Cantor's overall theory is understood, they cease to look like the products of the overactive imagination of a mathematical eccentric.

Cantor classified numbers with the same cardinality as belonging to the set "aleph null," or $\aleph_0$ ($\aleph$ is the first letter of the Hebrew alphabet). He called $\aleph_0$ a **transfinite number** (a number that is greater than any finite number). Amazingly, Cantor found that there are other transfinite numbers. These are sets of numbers with a greater cardinality than the integers. He labeled each successively larger transfinite number with increasing subscripts $\{\aleph_0, \aleph_1, \aleph_2, \ldots\}$.

Now, the reader may ask: how can there be different transfinite numbers? Cantor's proof is again remarkable for its simplicity. Suppose we take all the possible numbers that exist between 0 and 1 on the number line and lay them out in decimal form. Let's label each number $\{N_1, N_2, \ldots\}$. Note that there are so many possible numbers of the form $\frac{p}{q}$ between 0 and 1 that we could not possibly put them in any order. So, the numbers given here are just a sampling:

$$N_1 = .4225896\ldots$$
$$N_2 = .7166932\ldots$$
$$N_3 = .7796419\ldots$$
$$\ldots$$

How could we possibly construct a number that is not on that list? Here's how. Let's call it *C*. To create it, do the following: (1) for its first digit after the decimal point, choose a number that is greater by 1 than the first digit in the first place of N_1; (2) for its second digit, choose a number that is greater by 1 than the second number in the second place of N_2; (3) for its third digit, choose a number that is greater by 1 than the third number in the third place of N_3; (4) and so on:

$$N_1 = \underline{4}225896\ldots$$

The constructed number, *C*, would start with 5, rather than 4, after the decimal:

$$C = .5\ldots$$
$$N_2 = .7\underline{1}66932\ldots$$

The constructed number would have 2, rather than 1:

$$C = .52\ldots$$
$$N_3 = .77\underline{9}6419\ldots$$

The constructed number would have 0, rather than 9

$$C = .520\ldots$$
Etc.

Now, the number $C = .520\ldots$ is different from N_1, N_2, N_3, and so on, because its first digit is different from the first digit in N_1; its second digit is different from the second digit in N_2; its third digit is different from the third digit in N_3; and so on, ad infinitum. We have in fact just constructed a different transfinite number than $\aleph_0$. It appears nowhere in the previous list.

Reflections

The study of infinity is truly intriguing and mind-boggling and yet so simple. The Towers of Hanoi Puzzle can also be used in a simple way as an imaginary model of mathematical infinity. This can be done by imagining the number of pegs, their relative lengths, and the extension of the baseboard on which they are placed as having no limits. The game would thus go on ad infinitum.

There is something mystical about the concept of an eternal game—a concept found as well in *Das Glasperlenspiel* (1943, translated as *Magister Ludi*, 1949), the last novel written by the great German writer Hermann Hesse (1877–1962). In that novel, the meaning of life is revealed gradually to the master of a bead game—a game that involves making repeating patterns ad infinitum. There is no evidence to suggest that Hesse knew of Lucas's game, nor certainly that even if he did, it influenced him to write his masterpiece. But the idea that life is an eternal game played according to simple rules is captured brilliantly by both the puzzle and the novel. The appeal of Lucas's puzzle would seem to be as much metaphorical as it is mathematical.

Explorations

The Lucas and Ibn Kallikan Games

48. Here is a card version of the Towers of Hanoi game. Take four cards of the same suit, say, spades, in numerical order—an ace, a 2, a 3, and a 4. Put the cards in a space, calling it A. Set up two empty spaces, B and C, right next to the cards:

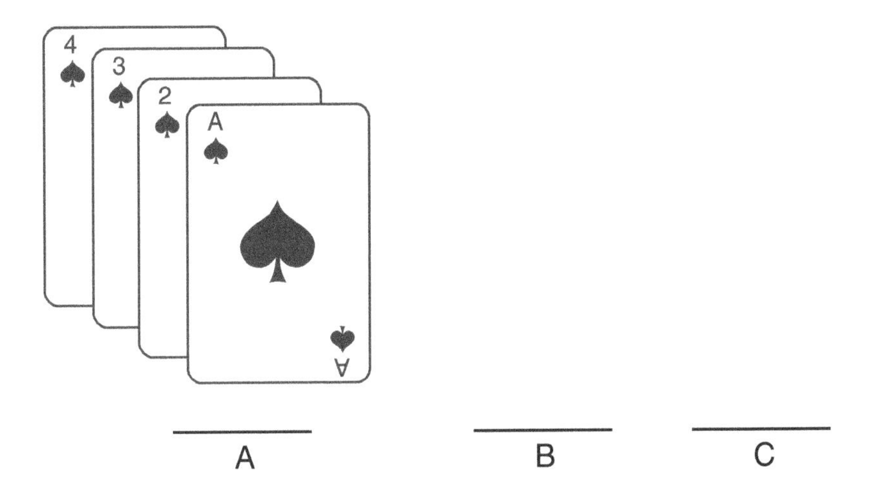

The object is to relocate the cards to space C in accordance with the same rules: (1) a larger-valued card may never be placed on top of a smaller-valued card; (2) only one card at a time can be moved to a new space.

49. Recall that the Mersenne formula ($2^n - 1$) generates primes when applied to certain squares on Ibn Kallikan's chessboard.

Square	Value of n	Value of Square	Mersenne Formula	Mersenne Prime
3rd	2	$2^2 = 4$	$(2^n - 1) = 2^2 - 1$	3
4th	3	$2^3 = 8$	$(2^n - 1) = 2^3 - 1$	7
6th	5	$2^5 = 32$	$(2^n - 1) = 2^5 - 1$	31
Etc.				

The values of n for the nine Mersenne primes on the chessboard are

Square	Value of n (in 2^n)
3rd	2
4th	3
6th	5
8th	7
14th	13
18th	17
20th	19
32nd	31
62nd	61

Do you detect a pattern?

50. What would happen if the conditions of Ibn Kallikan's puzzle were changed as follows?

- ▶ The number of grains on each even square is produced by multiplying the number of grains on the previous odd square by 2^n.
- ▶ The number of grains on each odd square is produced by halving the number of grains on the previous (even) square.

Again, we start with one grain on the first square. Do you detect any pattern to the sequence generated in this way?

51. The chessboard has been the source of all kinds of puzzles that explored mathematical ideas, ever since it came into use centuries ago. One that finds its way into virtually every puzzle anthology because of its apparent difficulty, yet deceptively simple solution, is the following:

> If two opposite corners of a checkerboard are removed, can the checkerboard be covered by dominoes? Assume that the size of each domino is the size of two adjacent squares of the checkerboard. The dominoes cannot be placed on top of each other and must lie flat.

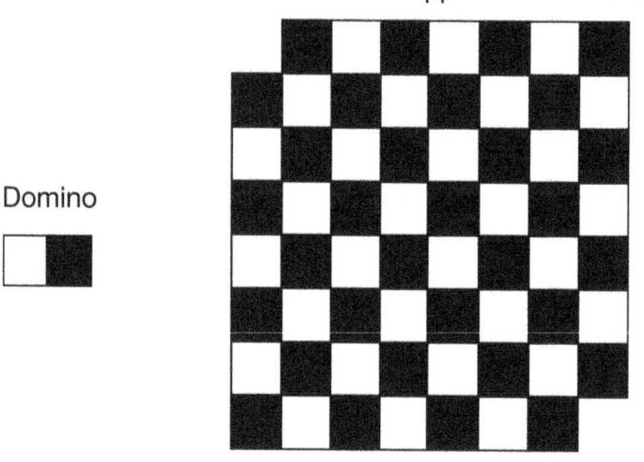

Infinity

52. Show that the whole numbers can be put in a one-to-one correspondence with

A. the set of multiples of 10

B. the subset of fractions with a constant numerator 1 and numerically ordered denominators, starting with 1 and increasing ad infinitum

53. Take the first transfinite number, $\aleph_0$:

A. What happens when you add 1 to it?

$\aleph_0 + 1 = ?$

B. What happens when you add any number, n, to it?

$\aleph_0 + n = ?$

C. What happens if you double it?

$\aleph_0 + \aleph_0 = 2\aleph_0 = ?$

Further Reading

Aczel, Amir D. *The Mystery of the Aleph: Mathematics, the Kabbalah and the Search for Infinity*. New York: Four Walls Eight Windows, 2000.

Kaplan, Robert, and Ellen Kaplan. *The Art of the Infinite: The Pleasures of Mathematics*. Oxford: Oxford University Press, 2003.

Lucas, François Edouard Anatole. *Récreations mathématiques*, 4 vols. Paris: Gauthier-Villars, 1882–1894.

Maor, Eli. *To Infinity and Beyond: A Cultural History of the Infinite*. Boston: Birkhäuser, 1987.

Rósza, P. *Playing with Infinity: Mathematical Explorations and Excursions*. New York: Dover, 1957.

Stewart, Ian. *From Here to Infinity: A Guide to Today's Mathematics*. Oxford: Oxford University Press, 1987.

Loyd's Get Off the Earth Puzzle

Whenever, therefore, people are deceived and form opinions wide of the truth, it is clear that the error has slid into their minds through the medium of certain resemblances to that truth.

SOCRATES (469–399 B.C.)

THE SHREWDEST PUZZLIST OF ALL TIME was, without question, the American engineer Sam Loyd. As we saw in chapter 4, he was the one who devised the Fifteen Puzzle, which turned out to be the first true worldwide "puzzle craze." Loyd invented a host of similarly clever puzzles and games that continue to baffle and entertain people to this day. Many of them suspend belief temporarily, producing the same mystifying effect that magic tricks do.

One of his "magic-inducing" puzzle gadgets, which never fails to bewilder people who come across it for the first time, is the Get Off the Earth

SAM LOYD (1841–1911)

Sam Loyd was born in Philadelphia. He studied engineering. However, after becoming the problem editor of the magazine *Chess Monthly* in 1860, he realized that he could make a comfortable living from puzzles alone.

Working out of a small, dusty office in New York City, Loyd produced over ten thousand puzzles in his lifetime. Most of them are extremely challenging, thus enticing "puzzle addicts" to spend countless hours trying to figure them out.

Puzzle. But like the Fifteen Puzzle, it, too, is not just an exercise in mental sleight of hand. As it turns out, the puzzle puts the spotlight on several important mathematical questions related to geometrical construction, and it is used by math teachers to emphasize to students the importance of examining all facts and all results, not just assuming them to be true. As such, it therefore qualifies as one of the ten greatest puzzles of all time.

The Puzzle

Loyd's puzzle is an ingenious "cut-and-slide" trick. The idea underlying its construction probably goes back to a puzzle included in a 1774 book titled *Rational Recreations*, by a certain William Hooper. Loyd created his version by fastening a smaller paper circle to a larger one with a pin so that it could spin around. Then, with appropriate artwork on both circles, he made the figure look like the earth, with thirteen Chinese warriors on it. Loyd patented his puzzle in 1897. It sold more than 10 million units:

When the smaller circle is turned slightly, shown as follows, the thirteen warriors turn mysteriously into twelve. Where did the thirteenth warrior go?

The Chinese warriors are made as assemblages of smaller pieces representing arms, legs, bodies, heads, and swords. When the earth is rotated, the pieces are rearranged in such a way that each Chinese warrior gains a sliver from his neighbor. For example, at the lower left, two warriors are next to each other. The top one is missing a foot. When the earth is rotated, he gains a foot from his neighbor on the right. That neighbor gains two feet (since he lost one) and one small piece of a leg. As a result of the rotation, one of the warriors will "lose" all his parts, making it seem that he has "disappeared."

To grasp the clever idea underlying the puzzle, consider a parallel vanishing trick. A rectangle, **ABCD**, that contains ten straight equidistant parallel lines within it is crossed by a dotted diagonal:

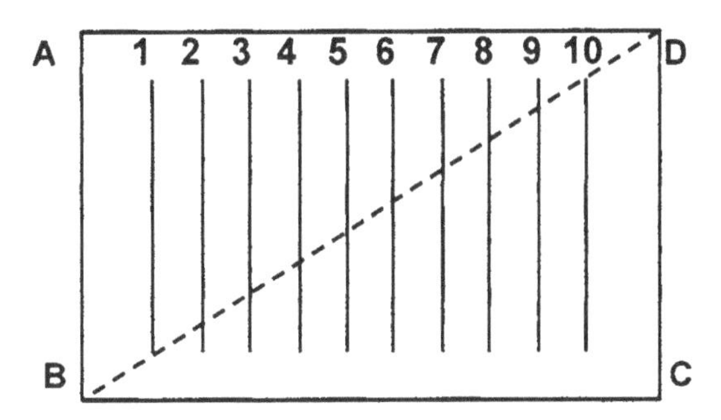

As can be seen, the diagonal touches the top point of line 10 and the bottom point of line 1. Readers should draw this rectangular figure on a piece of paper, making sure that the ten perpendicular lines are equal in length, parallel, and equidistant from one another. The dotted diagonal must be made to touch the top of line 10 and the bottom of line 1. Several drafts may be needed to come up with the correct figure. But it is crucial to do so; otherwise, the vanishing trick will not work.

We now cut the rectangle along the dotted line, producing an upper and a lower piece:

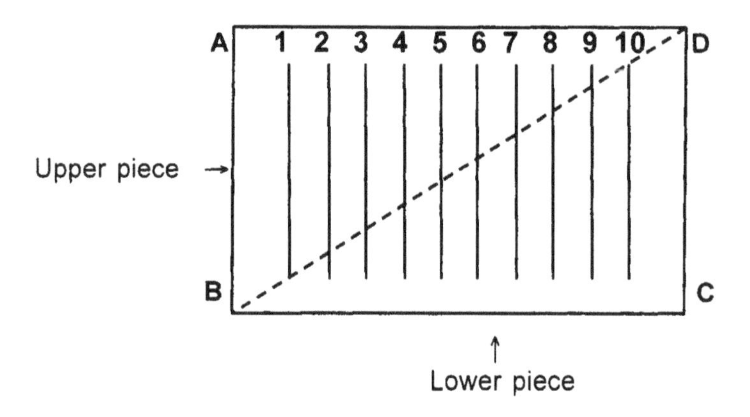

Next, we erase the numbers on the lines and the letters designating the rectangle. Then, we slide the lower piece down and to the left, only to the extent that the linear segments in the lower piece are "in sync" with the linear segments in the upper piece (see the following figure). In this way, we have seemingly preserved the internal lines in the rectangle. Two "protruding lines" are produced, however, as readers can confirm for themselves on their own paper versions:

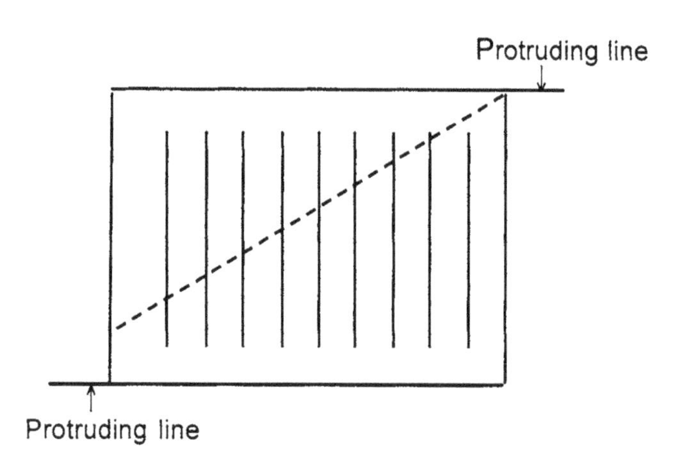

Let's cut out the two protruding lines. This produces a new and slightly smaller rectangular figure:

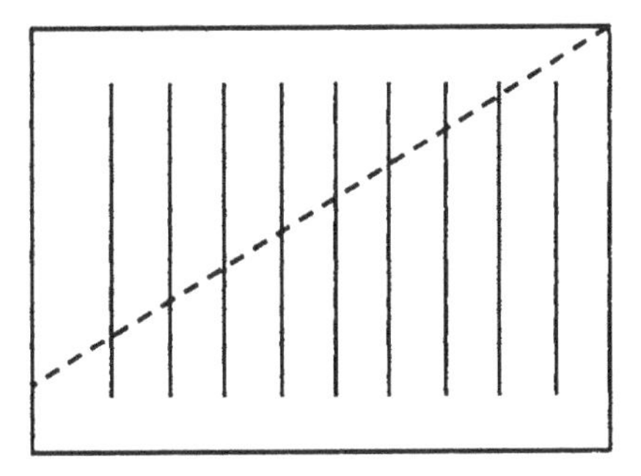

If we number the lines in the new figure and use letters to designate the new rectangle, we notice that there are now only nine internal lines in the rectangle:

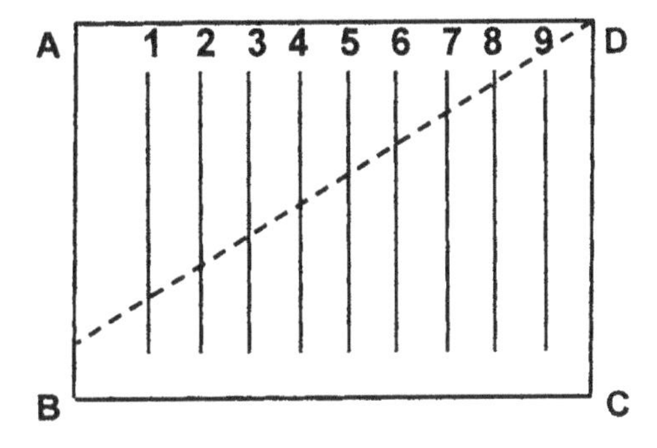

What happened to the tenth line? Nothing. Because of the slide, it has become coincident with side **DC** of the rectangle. It is, in effect, "hidden" by that side.

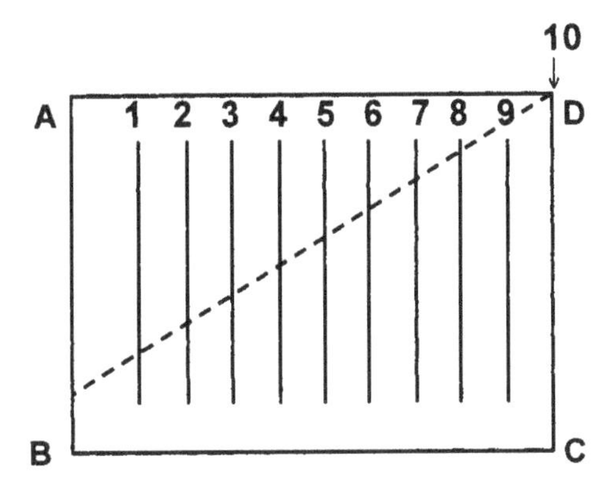

Let's analyze our "vanishing trick." Lines 1 and 10 remain the same after the cut, while the remaining eight lines (2 to 9) are sliced into two segments each. When we slide the lower piece, we produce new internal lines. Each one is now made up of its upper segment aligned with a lower segment that was previously part of the line to the immediate right (as we look at the diagram). The tenth line is still there, but it is now coincident with side **BC** of the new rectangle. Indeed, if we slide the lower part back up again, the tenth line will reappear.

This type of "cut-and-slide" trick was used by Loyd to create his mystifying Get Off the Earth Puzzle. When Loyd's smaller circle is turned, the body parts of the warriors, like the lines in our rectangle, are realigned, making it seem that one of the warriors, like our tenth line, has disappeared.

Mathematical Annotations

Constructing and dissecting figures with two instruments—the ruler (or straightedge) and the compass—have always constituted "concrete" techniques for investigating the properties of certain figures and for deducing theorems about them. As we saw here, Loyd's puzzle conceals within it a simple dissection technique, which never fails to stupefy people who are unaware of how Loyd used it to create his illusion.

Dissection

Loyd's puzzle really belongs to a genre of tricks based on dissection. Consider the following well-known puzzle. The original version appeared in 1868 (according to W. W. Rouse Ball, in Further Reading). Sam Loyd included it in his *Cyclopedia of Tricks and Puzzles* of 1914.

Start by dividing a square piece of paper into sixty-four smaller squares, as in a chessboard. The result is, of course, an 8×8 square:

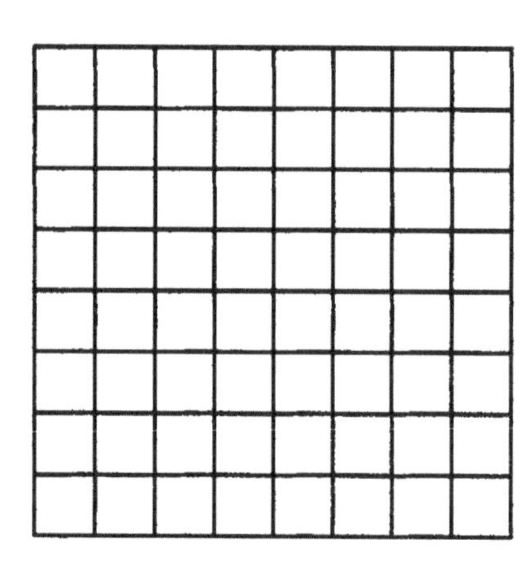

Cut up the chessboard into two trapezoids (figures 1 and 2) and two triangles (figures 3 and 4), as shown. A trapezoid is a four-sided plane figure with two opposite sides parallel to each other:

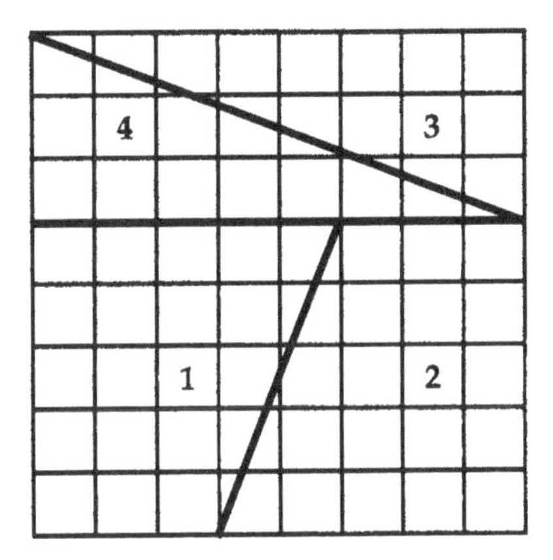

Finally, rearrange the four figures into a rectangle, shown as follows:

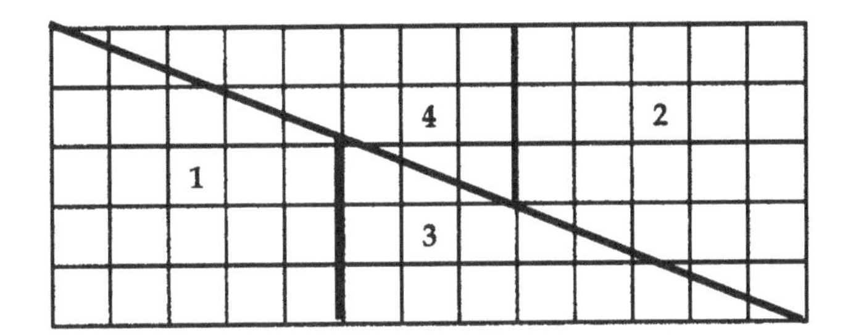

Now, count the small squares in the rectangle. There are 5×13, or 65, of them. But wait! That is one more than the sixty-four that were in the original square we used to make the rectangle! How did the extra small square get in there? The truth is that the edges of the four figures we cut out do not actually coincide along the diagonal. A close inspection will show that the diagonal is really a long and very narrow parallelogram that can barely be noticed. In the following drawing, it has been blackened to show where it is:

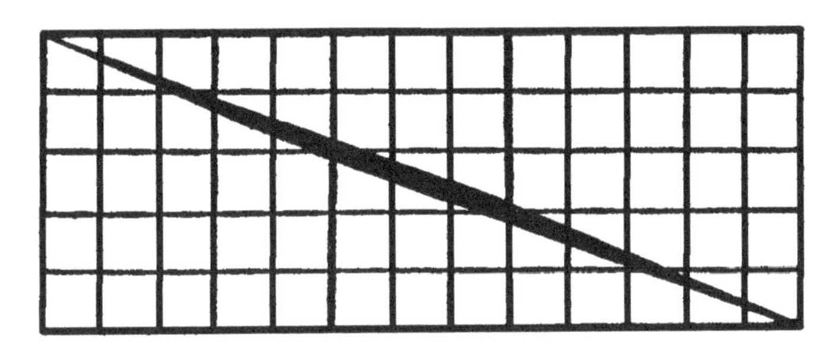

But this is not the end of the matter. If we subtract the area of the previous rectangle—$5 \times 13 = 65$—and subtract the area of the original square $8^2 = 64$, we get, of course, the difference of 1—which represents the area of the missing square. Let's write this out as follows:

Area of rectangle: 5×13

Area of original square: 8^2

Difference between the two areas: $(5 \times 13) - 8^2 = 1$

Now, look closely at the actual digits in the last expression. As it turns out, the digits—5, 8, and 13—are three consecutive numbers in the Fibonacci sequence (chapter 3)!

$$\{1, 1, 2, 3, \underline{5}, \underline{8}, \underline{13}, 21, 34, 55, 89, 144, 233, 377, 610, 987, \ldots\}.$$

And there is more. If we dissect squares of dimensions 3^2, 21^2, and 55^2 in the same way that we dissected our 8^2 square, we will produce rectangles of dimensions 5×2, 13×3, and 34×89 by rearrangement. Notice that all the digits in these expressions belong to the Fibonacci sequence. In each case, an extra little square unit is produced in the rearrangement process. Remarkably, subtracting the rectangles from the original squares in the same way that we did earlier also produces three consecutive Fibonacci numbers:

$5 \times 2 - 3^2 = 1 \rightarrow \ldots 2, 3, 5 \ldots$ (in the Fibonacci sequence)

$13 \times 34 - 21^2 = 1 \rightarrow \ldots 13, 21, 34 \ldots$ (in the Fibonacci sequence)

$34 \times 89 - 55^2 = 1 \rightarrow \ldots 34, 55, 89 \ldots$ (in the Fibonacci sequence)

Etc.

This result boggles the mind. It brings out once again that mathematics is all about the study of patterns, even if such study may, at times, have no practical application. The connection between Fibonacci numbers and a puzzle in dissection is one of those things that seems to lead nowhere but, nonetheless, seems to harbor a hidden significance that has not as yet been ascertained.

Dissection puzzles belong to the realm of the geometrical imagination. The term **geometry**—which derives from the ancient Greek *geo*, "earth," and *metrein*, "to measure"—describes what the early geometers did. They measured the size of fields, laid out accurate right angles for the corners of buildings, and calculated other practical things. And they used diagrams to represent their measurements and their layouts. More specifically, geometry is the branch of mathematics that deals with the construction, properties, and relationships of points, lines, angles, curves, shapes, and solids. Solving problems and puzzles in geometry requires, in fact, knowledge of how to draw and interpret diagrams correctly, as in the following classic puzzle:

Given the dimensions of the radius in inches, as shown in the following diagram, can you calculate the length of the rectangle's diagonal that goes from **A** to **B**?

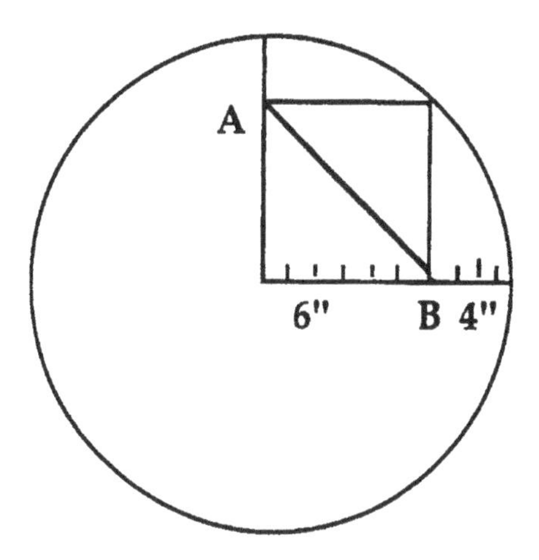

The puzzle seems to defy a solution. The insight comes from examining the diagram fully. Recall from high school geometry that the two diagonals of a rectangle equal each other in length. That, in fact, is the insight required to solve the puzzle. Go ahead and draw the other diagonal:

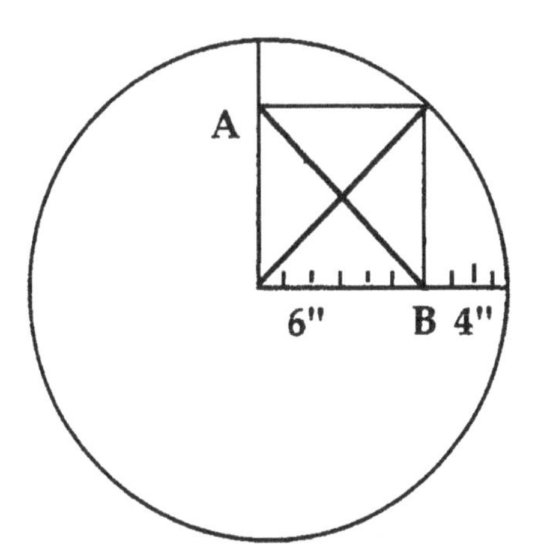

The diagonal that has just been drawn is, in effect, a radius of the circle. Since the latter is equal to 6 inches plus 4 inches, or 10 inches in total, and the radii of a circle are equal, the diagonal that was just drawn is also 10 inches long. And since the diagonals of a rectangle are equal, the length of the diagonal **AB** is thus 10 inches.

Optical Illusions, Ambiguous Figures, and Impossible Figures

Among other things, Loyd's puzzle constitutes a simple, albeit indirect, introduction to the world of optical illusions. These are figures that we interpret incorrectly. The topic of optical illusions is interesting psychologically and mathematically, having received extensive treatment in both fields. For the present purposes, it is sufficient to note that **optical illusions** trick the eye into interpreting some figure in an erroneous way. For instance, most people typically see the following line **AB** as longer than **CD**, even though the two are equal in length.

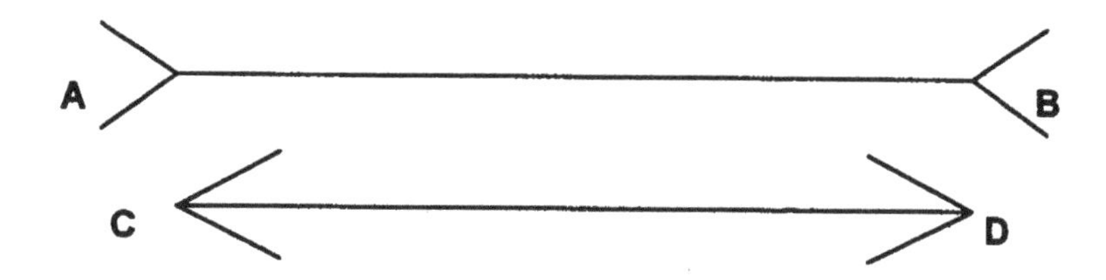

The illusion is called the Müller-Lyer Illusion, after the German physiologist Johannes Müller (1801–1858), who discovered it in 1840. The source of the illusion is, clearly, the different orientation of the two arrowheads. Readers should draw two equal parallel lines and then add arrowheads to them in a similar way, watching the "illusion effect" take place firsthand.

Following is another classic optical illusion, devised by the psychologist Johann Zöllner (1834–1882). The lines do not look parallel, but they are.

The lines appear slanted because the oblique little lines fool our eyes into interpreting them in that way. Again, readers should draw several equal vertical parallel lines and then add the slanted little lines to them in a similar way, watching the illusion effect take place firsthand.

Some figures are called **ambiguous** by psychologists, because they induce our eyes into perceiving them as something at one time, yet as something else at another. Perhaps the most famous of all ambiguous figures is the following one. It is found in virtually every introductory textbook on the psychology of perception:

We perceive the figure at one time as a vase and at another as the faces of two people looking at each other. Both of these perceptions come in flashes. Neither one can be maintained for very long. The illusion was devised by the Dutch psychologist Edgar Rubin around 1910. The cause of the ambiguity is, no doubt, the use of different shades. They produce a **chiaroscuro** effect, whereby at one time we cannot help but focus on the dark part of the figure and at another on the light part. *Chiaroscuro* is the term used by artists to refer to the distribution and the contrast of light and shade in a painting or a drawing. The term is derived from the Italian *chiaro* ("light") and *oscuro* ("dark") and generally refers to a technique that contrasts bright illumination with areas of dense shadow.

There is a third type of visual ruse worth mentioning here. It is known as an **impossible figure**. It is the product of a painting technique that dupes our eyes into viewing a two-dimensional drawing in a three-dimensional way—a technique developed by Renaissance artists such as

Filippo Brunelleschi (1377–1446) and Albrecht Dürer (1471–1528). The technique, known as **perspective drawing**, can be used, of course, to draw things like "cubes" on a two-dimensional surface, such as a piece of paper. But it can also be used cleverly to dupe our eyes into seeing figures as impossible. As an example, look at the following staircase:

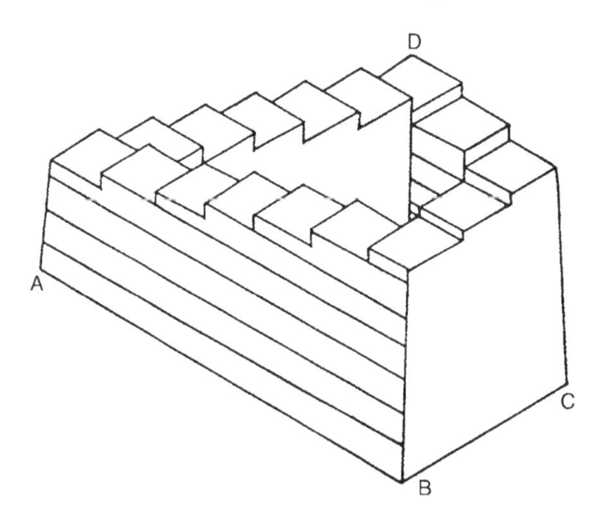

The staircase appears to be going up and down, defying common sense—it does not seem to have a highest or lowest step! If one starts "climbing" at D, moving counterclockwise, one ends up back at D, having apparently moved upward with each step and yet ending no higher than D. Similarly, if one moves clockwise, descending from D, one ends up, again, at D. The staircase thus appears to contradict all the principles of physics.

One of the most prolific producers of this kind of illusion is the Swedish artist and art historian Oscar Reutersvärd (1915–). His drawings have captured the attention of mathematicians and psychologists alike. Here is his "devil's triangle." It bears this name because it produces a jarring sense of distortion and surreal unease in the viewer. This particular version was actually created by the British biologist L. S. Penrose and his son the physicist Roger Penrose:

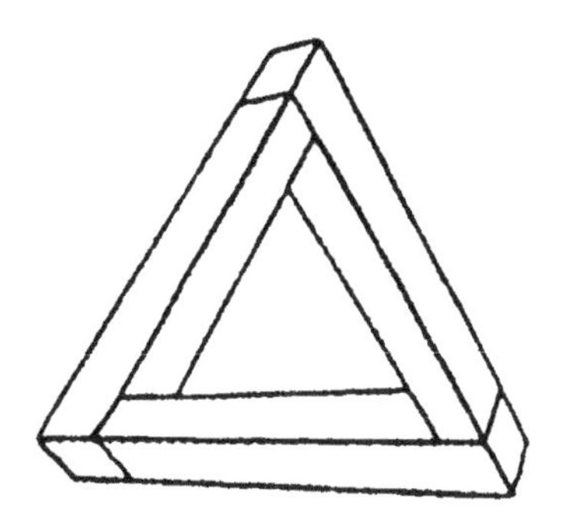

The artist who excelled at drawing such figures was Maurits Cornelis Escher (1898–1972). Escher's pictures explore the complex relationship between perception and representation. His interlocking figures; mirror images of cones, spheres, and cubes; connecting rings; and continuous spirals produce truly mind-boggling effects.

Reflections

Loyd's Get Off the Earth Puzzle warns us to be wary of how things appear on the surface. This was probably not the roguish Loyd's incentive for creating his money-making puzzle. But it turned out to be a perfect antidote for, and a way to counteract the naïve tendency of, accepting the evidence of the senses as accurate and trustworthy.

Optical illusions, ambiguous figures, and impossible figures also caution us to be careful about what our eyes tell us. They make it obvious that perception is not governed just by the physiology of vision. Rather, it is also the product of culturally based inferences operating at an unconscious level that we have learned to make about figures. These influence how our eyes interpret figures that are drawn on surfaces.

Explorations

Dissection and Rearrangement Puzzles

54. How can the following rectangle, with its two tabs, be cut into two pieces to make a complete rectangle?

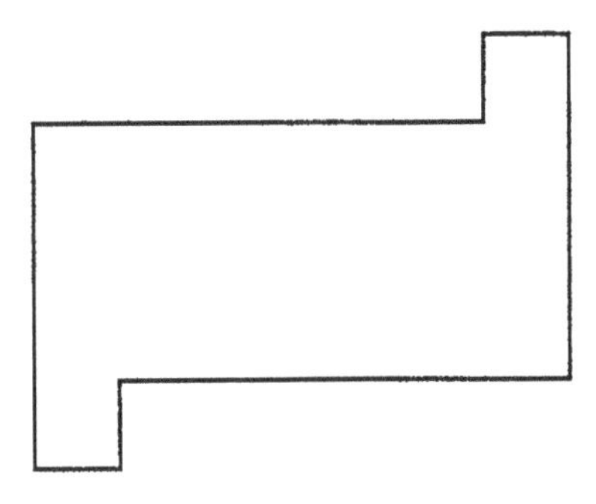

55. Look at the following figure:

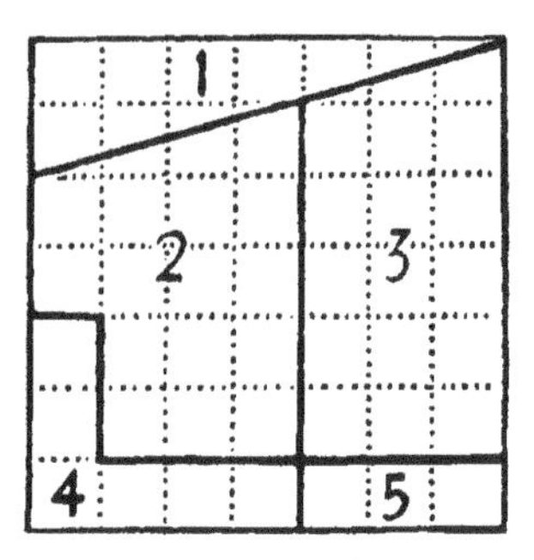

Readers can make it themselves by pasting a sheet of graph paper on a piece of cardboard, defining the boundary of the square as 7×7, and then drawing the internal lines as shown. Readers should then cut along the lines to make five pieces. When these are rearranged, in the following manner, a hole will appear in the center of the square!

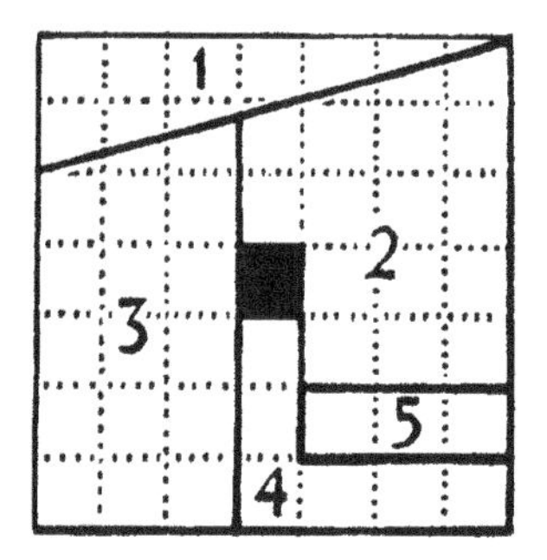

But that is not all. The original square had forty-nine smaller squares in it, whereas the square obtained through rearrangement of the parts has only forty-eight smaller squares. Which small square vanished and where did it go?

56. Look at the next drawing, in which there are six lightly colored pencils and seven darkly colored ones.

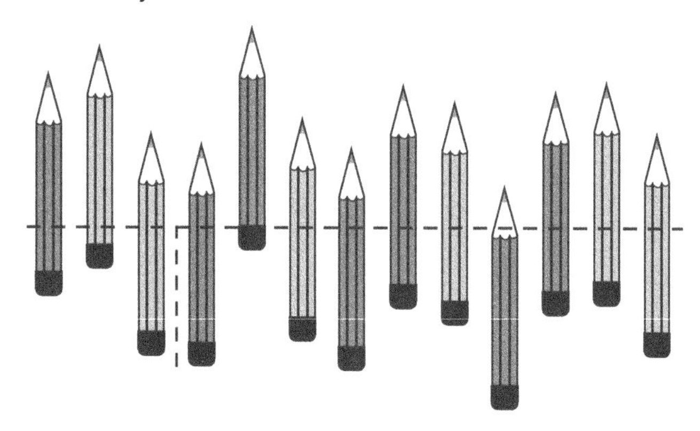

Now cut along the lines, reversing the lower left with the lower right parts. What happens?

Optical Illusions and Ambiguous Figures

57. Do you see two figures in the following drawing? What are they?

58. Which of the two pencils is longer? Measure them and find out.

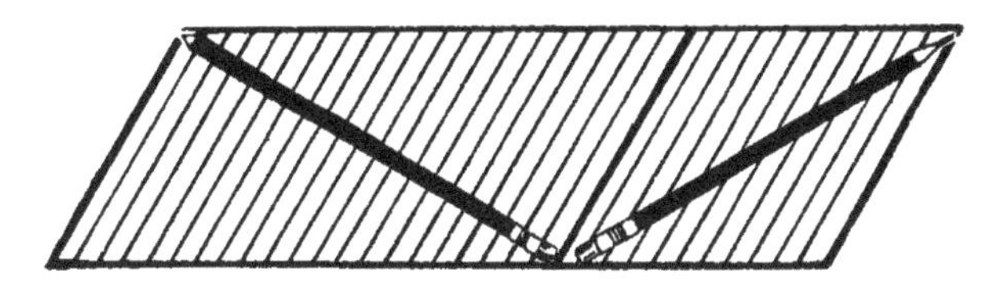

59. The following circle is made up of rings. One is not shaded. The radius of the large circle is 5, and the radius of each successive inner circle is 1, 2, 3, and 4. Which of the two shaded areas is larger?

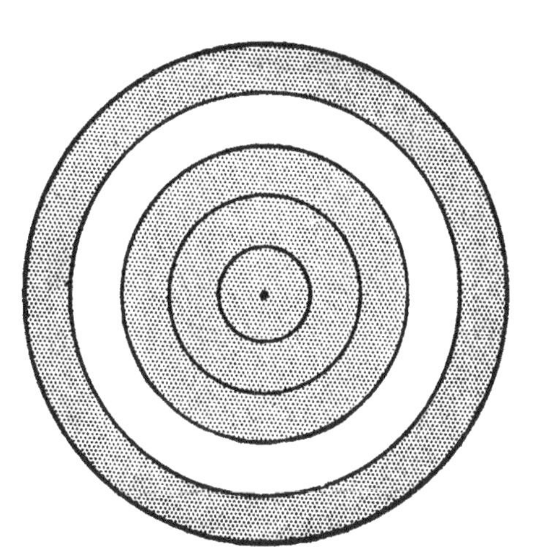

Further Reading

Ball, W. W. Rouse. *Mathematical Recreations and Essays*, 12th edition, revised by H. S. M. Coxeter. Toronto: University of Toronto Press, 1972.

Ernst, Bruno. *Impossible Worlds*. Köln, Germany: Taschen, 2002.

Gardner, Martin. *Entertaining Mathematical Puzzles*. New York: Dover, 1961.

Lindgren, H., and G. Frederickson. *Recreational Problems in Geometric Dissections and How to Solve Them*. New York: Dover, 1972.

Loyd, Sam. *Cyclopedia of Tricks and Puzzles*. New York: Dover, 1914.

———. *The Eighth Book of Tan*. New York: Dover, 1952.

———. *Mathematical Puzzles of Sam Loyd*, 2 volumes, compiled by M. Gardner. New York: Dover, 1959–1960.

Luckiesh, M. *Visual Illusions*. New York: Dover, 1965.

Rodgers, N. *Incredible Optical Illusions*. London: Quarto, 1998.

Shepard, R. N. *Mind Sights: Original Visual Illusions, Ambiguities, and Other Anomalies*. New York: W. H. Freeman, 1990.

Simon, S. *The Optical Illusion Book*. New York: William Morrow, 1984.

Epimenides' Liar Paradox

The way of paradoxes is the way of truth. To test Reality we must see it on the tight-rope. When the Verities become acrobats we can judge them.

OSCAR WILDE (1854–1900)

IN THE FIFTH CENTURY B.C., a host of intriguing debates broke out in Greece over the nature and the function of logic in science and mathematics. Prominent participants were the philosopher Parmenides (c. 510 B.C.) and his disciple Zeno of Elea. The latter became famous for a series of clever arguments that seemed to defy common sense. The arguments came to be known as **paradoxes** (meaning, literally, "conflicting with expectation").

A group of traveling teachers, called the Sophists (from Greek *sophos*, "clever"), sided with Zeno, arguing that paradoxes exposed logical thinking as essentially deceptive. The great philosopher Aristotle (384–322 B.C.),

ZENO OF ELEA (c. 489–435 B.C.)

Little is known of Zeno's life, other than that he lived in the Greek colony of Elea in southern Italy.

With his ingenious paradoxes, Zeno showed that a purely logical approach to a description of reality would force us to conclude that motion is impossible. Despite their iconoclastic intent, the ideas built into the paradoxes led gradually to the establishment of the calculus and to a reconsideration of the logical foundations of mathematics.

on the other hand, dismissed Zeno's paradoxes as exercises in specious reasoning. The central characteristic of the human mind, Aristotle insisted, was its ability to think logically. He then proceeded to give logic a formal structure, called syllogistic.

Here is an example of an Aristotelian **syllogism**:

Major premise: All humans are mortal.

Minor premise: Socrates is human.

Conclusion: Therefore, Socrates is mortal.

The major premise states that a category has (or does not have) a certain characteristic and the minor premise says that a certain thing is (or is not) a member of that category. The conclusion then affirms (or negates) that the thing in question has that characteristic. As clever as they were, Aristotle asserted, paradoxes were ultimately inconsequential because they did not impugn the validity of the syllogism. But Aristotle's response was not the end of the matter. On the contrary, the history of logic and mathematics recounts a poetic vindication of Zeno's stand.

Paradoxes are essentially puzzles in logic. The story goes that during the debates, Protagoras (c. 480–411 B.C.) concocted one of the most vexing of all paradoxes. Protagoras was the first philosopher to call himself a Sophist. The paradox has come to be known as the Liar Paradox. Its most famous articulation has, however, been attributed to a Cretan named Epimenides in the sixth century B.C. Almost nothing is known about his life, other than the fact that he was a celebrated poet and a prophet of Crete. The Liar Paradox belongs on the list of the top ten greatest puzzles of all time, not only because it continues to astound people to this day but also because it has had many repercussions for the study of logic. Some of these will be discussed in this chapter.

The Puzzle

Before dealing with the Liar Paradox, it is useful to briefly consider a similar kind of paradox that virtually everyone knows:

Which came first, the chicken or the egg?

If you said that the chicken came first, then someone could counter that such a thing would be impossible because the chicken had to hatch first from an egg. If you say that the egg came first, then someone could again counter by saying that such a thing would also be impossible because the egg had to be laid first by a chicken. The question of which came first—the chicken or the egg—seems intractable. Answering it only produces an exchange that will go around and around in circles forever!

The Liar Paradox evokes the exact same kind of "circularity." It has come down to us more or less in the following form:

> The Cretan philosopher Epimenides once said: "All Cretans are liars." Did Epimenides speak the truth?

Let's assume that Epimenides spoke the truth. Thus, his statement "All Cretans are liars" is a true statement. However, from this, we must deduce that Epimenides, being a Cretan, is also a liar. But this is a contradiction. Obviously, we must discard our assumption. Let's assume the opposite—namely, that Epimenides is in fact a liar. But, then, if he is a liar, the statement he just made—"All Cretans are liars"—is true. But this is again a contradiction—liars do not make true statements. Obviously, we are confronted with a circularity, not unlike that of the chicken and the egg.

The British mathematician P. E. B. Jourdain invented an interesting version of the Liar Paradox in 1913, which brings out its essential nature in a concrete way:

> The following is printed on one side of a card: "The statement on the other side of this card is true." But on the card's other side, the statement reads: "The statement on the other side of this card is false." What do you make of the card?

The card makes you go back and forth, from one side to the other, scratching your head. The reader may, at this point, wonder what the Liar Paradox has to do with mathematics. The answer is that mathematics has always been thought to be free of logical circularity. But it is not. This is why the Liar Paradox has fascinated mathematicians throughout history, becoming, over time, one in a series of clever paradoxes that has brought about revolutionary changes in mathematics.

Mathematical Annotations

The dream of mathematicians has always been to provide a firm logical foundation to mathematics that would be free of circularity. But paradoxes have always stood in the way of this master plan. They expose logic as problematic. For this reason they have, paradoxically (no pun intended), played a crucial role in the history of mathematics. Attempts to resolve the issues that they raise have, in fact, led to significant debates and subsequent discoveries and developments.

Undecidability

The source of the circularity in the Liar Paradox is, of course, the fact that it was Epimenides, a Cretan, who made the statement "All Cretans are liars." It is an example of the logical problems that arise from **self-referentiality**. This refers to the fact that the maker of a statement includes himself or herself in the statement. The English philosopher Bertrand Russell (1872–1970) found the paradox to be especially troubling, feeling that it threatened the very foundations of logic and mathematics.

To examine the nature of self-referentiality more precisely, Russell formulated his own version of the Liar Paradox, called the Barber Paradox:

> The village barber shaves all and only those villagers who do not shave themselves. So, shall he shave himself?

The barber is "damned if he does and damned if he doesn't," as the colloquial expression goes. Let's say he decides to shave himself. He would end up being shaved, of course, but the person he would have shaved is himself. And that contravenes the requirement that the village barber shave "all and only those villagers who do not shave themselves." The barber has, in effect, just shaved someone who shaves himself! So, let's assume that the barber decides not to shave himself. But, then, he would end up being an unshaved villager. Again, this goes contrary to the stipulation that he, the barber, shave "all and only those villagers who do not shave themselves"—including himself! It is not possible, therefore, for the barber to decide whether or not to shave himself. Russell argued that such "undecidability" arises because the barber is himself a member of the village. If the barber were from a different village, the paradox would not arise.

Like the German philosopher Gottlob Frege (1848–1925), Russell sought to find a system of logical argumentation that would exclude self-referentiality. Using a notion developed two millennia earlier by Chrysippus of Soli (c. 280–206 B.C.), Frege claimed that circularity could be avoided from statements such as the Liar Paradox by considering their *form* separately from their *content*. In this way, one could examine the consistency of statements, known more technically in logic as **propositions**, without having them correspond to anything (such as barbers, villages, Cretans, etc.). Frege's approach was developed further by the Cambridge logician Ludwig Wittgenstein (1889–1951), who used symbols, rather than words, to ensure that the form of a proposition could be examined in itself for logical consistency, separate from any content to which it could be applied. If the statement "It is raining" is represented by the symbol p and the statement "It is sunny" by q, then the proposition "It is either raining or it is sunny" can be assigned the general symbolic form $p \vee q$ (with $\vee$ = "or"). A proposition in which the quantifier "all" occurs would be shown with an inverted $\forall$. So, the statement "All Cretans are liars" would be represented as $\forall p$. If the form held up to logical scrutiny, then that was the end of the matter. The problem, Wittgenstein affirmed, is that we expect logic to interpret reality for us. But that is expecting way too much from it. Wittgenstein's system came to be known as "symbolic logic"—a system of representation prefigured by none other than the puzzlist Lewis Carroll in his ingenious book *The Game of Logic* (reprinted in 1958).

Russell joined forces with Alfred North Whitehead (1861–1947) to produce a system of symbolic logic called the *Principia Mathematica* ("The Principles of Mathematics") in 1913. The objective of the two philosophers was to solve the problem of undecidability, such as the one faced by the village barber, by separating the form of propositions completely from their reference to "real-world" content. At first thought, Russell and Whitehead's proposal would seem to make a lot of sense. After all, in music the form of a melody is all that really counts, no matter what emotions it may evoke. As long as it is consistent with harmonic practices, it is a valid melody. But, alas, it became obvious after publication of the *Principia Mathematica* that the forms of the propositions themselves led to unexpected problems. To solve these, Russell introduced the notion of "types," whereby certain types of propositions would be classified into different levels (more and more abstract) and thus considered separately from other types. This seemed to avoid the problems—for a while, anyhow.

The Polish mathematician Alfred Tarski (1902–1983) developed Russell's idea of types further by naming each increasing level of abstract statements a metalanguage. A **metalanguage** is, essentially, a statement about another statement. At the bottom of the metalanguage hierarchy are straightforward statements about things, such as "Earth has one moon." Now, if you say "The statement that earth has one moon is true," you are using a different type of language, because it constitutes a statement about a previous statement. It is a metalanguage. The problem with this whole approach is, of course, that more and more abstract metalanguages are needed to evaluate lower-level statements. And this can go on ad infinitum. In effect, Tarski's system only postpones making final decisions about "what is what." Consider, for example, the following two statements:

1. The next sentence is false.
2. The previous sentence is true.

Statement 1 refers to another statement (2). So, it belongs to some metalanguage. Now, statement 2, which is the target of statement 1, also says something about statement 1. So, it, too, belongs to some metalanguage. But, wait! This means that it belongs to two levels at once! It would seem that Tarski's system itself produces self-referentiality among metalanguages!

This whole line of investigation was finally (and mercifully, some would claim) brought to an abrupt end in 1931 by the German logician Kurt Gödel (1906–1978). While at Princeton, Gödel showed why self-referentiality is a fact of human life, no matter how hard we try to eliminate it from our logical systems. Before Gödel, it was taken for granted that every proposition within a logical system could be either proved or disproved within that system. But Gödel startled the academic world by showing that this was not the case! He argued that a logical system invariably contains a proposition within it that is "true" but "unprovable." Gödel's argument is far too technical to discuss here in an in-depth manner. For our present purposes, it can be paraphrased as follows:

> Consider a mathematical system T that is both correct—in the sense that no false statement is provable in it—and that contains a statement, S, that asserts its own unprovability in the system. S can be formulated simply as: "I am not provable in system T." What is the truth status of S? If it is false, then its opposite is true, which means that S is provable in system T. But this goes contrary to our assumption that no false statement is provable in the system. Therefore, we conclude that S must be true, from which it follows that S is unprovable in T, as S asserts. Thus, S is true but not provable in the system.

The American puzzlist Raymond Smullyan (see Smullyan's 1997 book, listed in Further Reading) provides a clever puzzle version of Gödel's argument, as follows:

> Let us define a logician to be accurate if everything he can prove is true; he never proves anything false.
>
> One day, an accurate logician visited the Island of Knights and Knaves, in which each inhabitant is either a knight or a knave, and knights make only true statements and knaves make only false ones. The logician met a native who made a statement from which it follows that the native must be a knight, but the logician can never prove that he is!
>
> What was the statement?

The "Gödelian" statement is: *You cannot prove that I am a knight.* The speaker is either a knave or a knight. Let's assume that he is a mendacious knave. In that case, the statement would, of course, be false. Its opposite would be true—namely, *You can prove that I am a knight.* But the speaker is not a knight; he is a knave. Since the puzzle asserts that an accurate logician is incapable of proving anything false, he cannot prove therefore that the native is a knight in this case as the mendacious speaker asserted!

Therefore, the speaker is not a knave. Let's assume that he is a knight instead. This means that the statement *You cannot prove that I am a knight* is true, since our speaker is now a truthful person. But if it is true, then the logician cannot prove again that the native is a knight—the statement declares as much. So, even though the native is a knight, the logician will never be able to prove it!

Gödel's demonstration showed, once and for all, that logical systems are "faulty," because they invariably contain a statement ("I am not provable") that is *undecidable* in them. Its repercussions are being felt throughout mathematics and philosophy to this day. Perhaps the most appropriate way to eliminate self-referentiality is to simply outlaw it. The "technique" of prohibition has, in fact, already been used by mathematicians with regard to division by 0. Why? Because it would lead to contradictory results (see **contradiction**, in the glossary), as the following demonstration shows:

1. Assume that $a = b$.
2. Multiply both sides of the equation by a: $a^2 = ab$.
3. Subtract b^2 from both sides: $a^2 - b^2 = ab - b^2$.
4. Factor both sides: $(a + b)(a - b) = b(a - b)$.
5. Divide both sides by $(a - b)$: $a + b = b$.
6. Since $a = b$, from (1), (5) can be rewritten as: $b + b = b$.
7. Therefore: $2b = b$.
8. Dividing both sides by the common factor b, we get: $2 = 1$.

We have thus proven that $2 = 1$, or have we?

This anomaly arises because we started off by assuming that $a = b$, which means that $(a - b) = 0$:

$$a = b.$$

Subtract b from both sides:

$$(a - b) = (b - b)$$
$$(a - b) = 0.$$

Thus, when we divided the equation $(a + b)(a - b) = b(a - b)$ by $(a - b)$, we were in effect dividing it by 0. The reason for prohibiting division by 0 is, clearly, a practical one—it is better to retain a system of numeration that has proven to be highly useful in everyday life, than to throw it completely out because one of the numbers within it is problematic. Mathematical life goes on without division by zero. So, too, logical life will go on without self-referential statements.

In their 1986 book *The Liar* (see Further Reading), a mathematician named Jon Barwise and a philosopher named John Etchemendy adopted a similar "practical" view of the Liar Paradox. As they assert, the paradox arises only because we allow it to arise. When Epimenides says, "All Cretans are liars," he may be doing so simply to confound his interlocutors. His statement may also be the result of a slip of the tongue. Whatever the case, the intent of Epimenides' statement can be determined only by assessing the context in which it was uttered, along with Epimenides' reasons for saying it. Once such factors are determined, no paradox arises.

Incidentally, Gödel's basic argument may shed light on why some problems may not be solvable with logic alone. They may, in effect, be "undecidable" within our systems of logic. We have encountered one of these in the Four-Color Problem, discussed in chapter 5. Another is Goldbach's Conjecture. In a letter to Euler in 1742, the mathematician Christian Goldbach (1690–1764) conjectured that every even integer greater than 2 could be written as a sum of two primes:

$4 = 2 + 2$

$6 = 3 + 3$

$8 = 5 + 3$

$10 = 7 + 3$

$12 = 7 + 5$

$14 = 11 + 3$

$16 = 11 + 5$

$18 = 11 + 7$

Etc.

No exception is known to Goldbach's Conjecture, but there still is no valid proof of it. Goldbach also conjectured that any number greater than 5 could be written as the sum of three primes:

$6 = 2 + 2 + 2$

$8 = 2 + 3 + 3$

$7 = 2 + 2 + 3$

$9 = 3 + 3 + 3$

$10 = 2 + 3 + 5$

$11 = 3 + 3 + 5$

Etc.

Maybe the conjecture is one of those things that is undecidable within our systems of logic. From a practical perspective, an "explanation" for the conjecture may be unnecessary anyhow, for it would probably not change the world in any significant way. But, for some reason, we continue to search for a proof, as if impelled by our ancient Theban Sphinx to do so—no matter what the cost!

Limits

Paradoxes have not only had a great impact on the study of logic, they have also led to the crystallization of a host of mathematical notions. One of these is the notion of **limits** (the boundary number or point that is approached by a function), which can be traced back to Zeno's famous paradox of the runner, by which he argued that a runner would never be able to reach a finish line, if one used *logical* argumentation. He argued his case as follows. The runner must first traverse *half* the distance to the finish line. Then, from mid-position, the runner would face a new, but similar, task—he must traverse *half of the remaining distance* between himself and the finish line. But from the new position, the runner would face a similar task—he must once more cover *half of the new remaining distance* between himself and the finish line. Although the successive half distances between himself and the finish line would become increasingly (indeed, infinitesimally) small, the wily Zeno concluded that the runner would come very close to the finish line but would never cross it. The successive distances that the runner must cover form an infinite geometric series, each term of which is half of the one before: $\{\frac{1}{2}, \frac{1}{4}, \frac{1}{8}, \frac{1}{16}, \ldots\}$. The sum of the terms in this sequence will never reach 1, the whole distance to be covered:

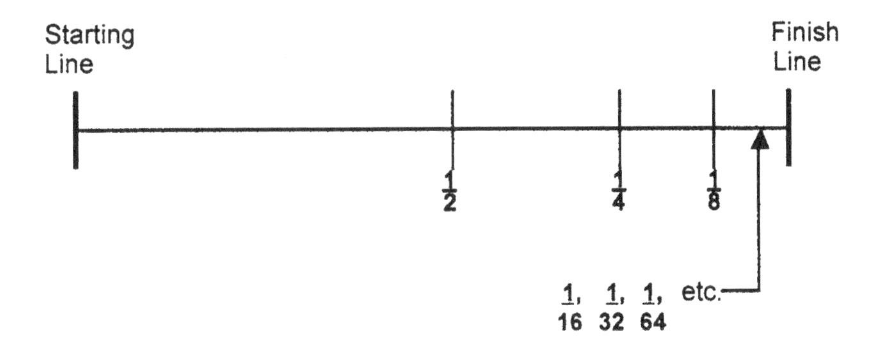

The English scientist Sir Isaac Newton (1642–1727) and the German philosopher and mathematician Gottfried Wilhelm Leibniz (1646–1716) were probably contemplating Zeno's runner paradox when they came up, independently, with an ingenious, yet remarkably simple, solution to it. They simply asserted that the sum to which the series $\{\frac{1}{2}, \frac{1}{4}, \frac{1}{8}, \frac{1}{16}, \ldots\}$ converges as it approaches infinity is the distance between the starting line and the finish line. Thus, the *limit* of the runner's movement is, in fact, the unit distance of 1. This notion became the basis for establishing a new branch of mathematics, known as the **calculus**, which is concerned with such concepts as the rate of change, the slope of a curve at a particular point, and the calculation of an area bounded by curves. It is beyond the scope of this book to discuss the historical roots of calculus. Suffice it to say, calculus led to a radical reconsideration of philosophical and mathematical ideas about the world. Indeed, when the calculus was first proposed, it met with acerbic criticism from philosophers and religious leaders. The Irish prelate and philosopher George Berkeley (1685–1753), for instance, charged that it was a useless science because it dealt with small, meaningless quantities. But the calculus easily survived such attacks, for the simple reason that it provided a powerful conceptual framework for answering the classical unsolved problems of physics and the paradoxes of Zeno.

The idea of limits was not unknown before Newton and Leibniz. In an ancient Egyptian manuscript titled the *Ahmes Papyrus*, after the Egyptian scribe Ahmes who copied it, or the *Rhind Papyrus*, after the Scottish lawyer and antiquarian A. Henry Rhind (1833–1863), who purchased it in 1858 while vacationing in Egypt, one finds this very idea. It is used to estimate the value of π, or pi. The original manuscript was written nearly four thousand years ago, during the same period in which another famous document of Egyptian mathematics, the *Moscow Papyrus* (named after its current location), was written. The papyrus contains eighty-four challenging mathematical problems. Its estimation of the value of π—the ratio of the circumference of a circle to its diameter, which is approximately 22/7 or, to five decimal places, 3.14159—is found in problem 48:

> What is the area of a circle inscribed in a square that is 9 units on its side?

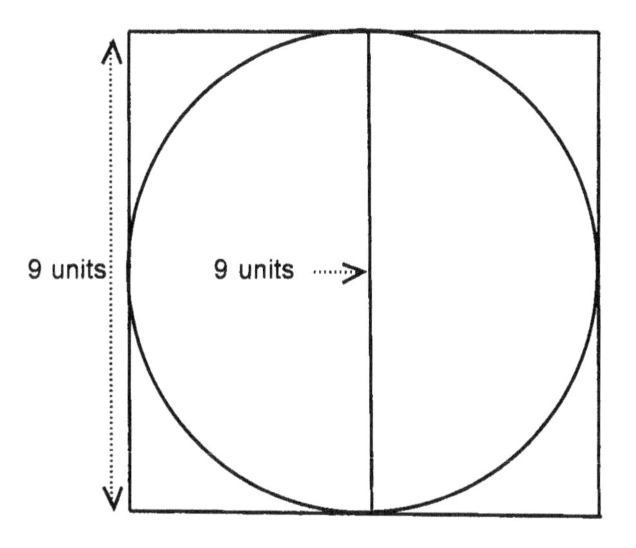

Notice that the diameter of the circle is also 9 units, as shown in the diagram. The clever Ahmes (or whoever the real author of the *Rhind Papyrus* was) solved it by a method that foreshadowed the technique of limits. He asked, essentially, What if the circle is transformed into a polygon? He then proceeded to do exactly that by trisecting each side of the square, as shown in the following diagram, thus producing nine smaller squares within it (each 3×3). He also drew the diagonals in the corner squares, as shown. With such modifications to the diagram, Ahmes produced an octagon, which he assumed to be close enough in area to the circle for the practical purposes of the problem at hand:

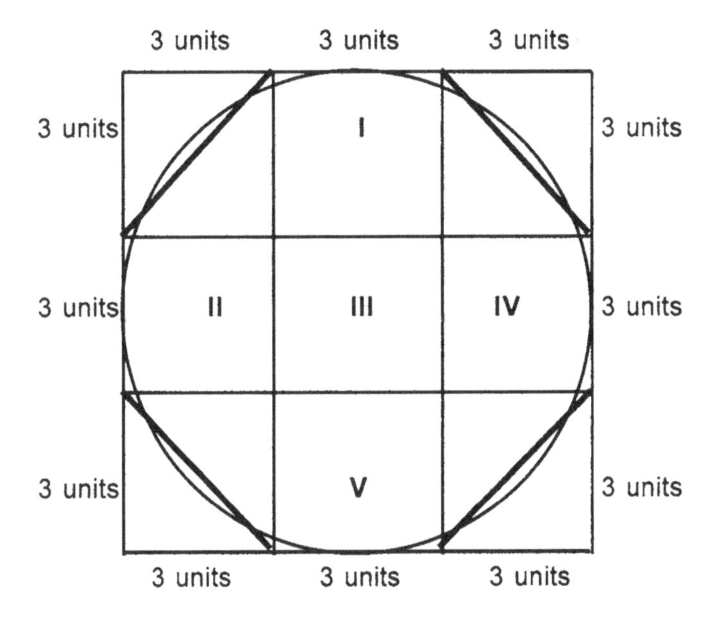

Now, the area of the octagon can be computed easily, because it is made up of seven smaller squares (all equal in area)—namely, the five inner squares (I, II, III, IV, V), plus half the four small corner squares, which is equivalent, of course, to two squares. The area of one small square is $3 \times 3 = 9$ square units. The total area of seven such squares is, therefore, $9 \times 7 = 63$ square units. With a bit of convenient cheating, the resourceful Ahmes assumed the octagon's area, and hence the circle's area, to be 64. He then estimated the value of π as follows. Recall that the diameter of the circle is 9:

Area of circle: $\pi r^2 = 64$

Diameter: 9

Radius (r): $\frac{9}{2}$

So, $r^2 = \left(\frac{9}{2}\right)^2 = 20.25$

Thus, since $\pi r^2 = 64$, and $r^2 = 20.25$: $20.25\pi = 64$. Dividing both sides by 20.25: $\pi = 64/20.25$. This is equivalent to: $\pi = 3.16049\ldots$

A strikingly similar insight was used over one thousand years later by the great Sicilian mathematician Archimedes (c. 287–212 B.C.). Archimedes inscribed a regular polygon in a circle. The difference between the perimeter of a polygon and the circumference of a circle, he argued, could be made as small as one desired by progressively increasing the number of sides of the polygon. The limiting figure of such an incremental procedure was the circle, and the limiting area of the "infinite-sided" polygon, therefore, the area of the circle. One will never be able to calculate the circle's area exactly, Archimedes observed, but one can clearly approximate it as accurately as one wishes. From this approximation, one can thus deduce a value for π. A world in which π is not known is, of course, conceivable. But what we now know about circular objects in the world, like the sun and the tides, would be much more rudimentary. Our ability to describe natural phenomena would be reduced to rudimentary conceptual dimensions.

FUNCTIONS

The term **function** is used in mathematics to indicate the relationship between two or more variables.

For the function $y = f(n)$ (read as "y is a function of n"), let's assume that $f(n) = 1/n$. Thus: $y = f(n) = 1/n$.

Now, we can always determine a value for y by simply specifying a value for n.

Examples:

When $n = 2$,

$$y = f(n) = \frac{1}{n} = \frac{1}{2}.$$

When $n = 3$,

$$y = f(n) = \frac{1}{n} = \frac{1}{3}.$$

When $n = 1{,}000{,}000$,

$$y = f(n) = \frac{1}{n} = \frac{1}{1{,}000{,}000}.$$

Because the value of y depends on the value of $f(n)$, y is called the **dependent variable** and $f(n)$ the **independent variable**.

The value of π, to which Archimedes' inscribed polygon calculations converged, is a perfect example of a limit. One can approach the limit of $\pi = 3.14159\ldots$ closer and closer by increasing the number of polygons ad infinitum.

Limits indicate the ways in which functions behave (see the sidebar). The following formula says in symbols that the limit of the function $(1/n)$ approaches 0 as n gets to be bigger and bigger:

$$\lim_{n \to \infty} (1/n) = 0.$$

The previous formula is read as follows: "The limit of the function $(1/n)$ as n approaches infinity ($n \to \infty$)—that is, as it gets bigger and bigger—is 0."

The calculus is, in effect, a way of computing limits as measures of changing events (speeds, movements, etc.): How fast does a stone fall two seconds after it has been dropped from a cliff? What is its speed at any point in time? It tries to find a quantity by figuring out the rate at which it is changing.

Reflections

Despite all the radical implications it has had for the development of mathematics, the Liar Paradox ultimately does not invalidate the use of logic in everyday life. In our three-dimensional world, if it is true, for instance, that building A is higher than building B, and that building C is higher than building A, then we can conclude, without any shadow of a doubt, that building C is higher than building B. Nevertheless, the Liar Paradox continues to warn us against believing that logic is the only path to knowledge. Hunches and experience are probably just as important, if not more so, for grasping the meaning of things.

Incidentally, it is relevant to note that the very concept of "logic" itself does not originate in the world of mathematics but in a more mystical domain. It was in sixth-century B.C. Greece that the philosopher Heraclitus asserted that the world was governed by the Logos, a divine force that produces order in the flux of nature. Shortly thereafter, **Logos** came to be viewed as a rational divine power that directed the universe. Through the faculty of reason, all human beings were thought to share in it. Even the Gospel according to John identifies Logos ("the Word") as a spiritual force: "In the beginning was the Word, and the Word was with God." Only much later was Logos viewed to be the power of human intellect to reason things out.

Explorations

Logic

The following puzzles are designed to allow the reader to directly explore the nature of logical thinking. All are versions of classic logic puzzles and paradoxes.

60. The following is written on a piece of paper: "This sentence is false." Is the sentence true?

61. A gold coin is in one of the following three boxes, each of which has an inscription written on it as follows:

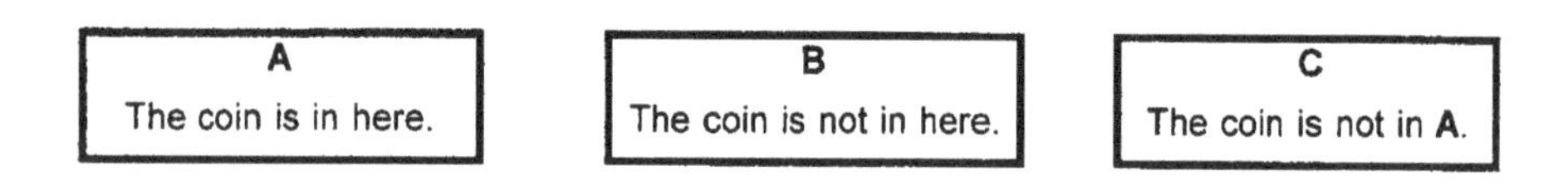

Can you tell where the coin is if, *at most*, only one of the inscriptions is true?

62. A jewelry box bears the following inscription:

This box was not made by a truth-teller

Did a truth-teller or a liar make the box?

63. Let's assume that $x + y = y$. Now, let's assign some values to x and y as follows.

If $x = 0$ and $y = 1$, we get:

$x + y = y$

$0 + 1 = 1$, which is, of course, correct.

If $x = 1$ and $y = 2$, we get:

$x + y = y$

$1 + 2 = 2$, which is incorrect.

How is this so?

64. Is the following sentence true? "This sentence has seven words." What is its negative version? Is that true?

65. A man is looking at a photo: "Brothers and sisters have I none, but this man's son is my father's son." Who is the person in the photo?

66. The first customer in a Milwaukee bookstore gave the salesclerk a $10 bill for a $3 book. The salesclerk, having no change, took the $10 bill across the street to a record store to get it broken down into ten $1 bills. The salesclerk then gave the customer the book worth $3 and seven $1 bills as change.

An hour later the record-store salesclerk returned the $10 bill and demanded her money back, claiming that the bill was counterfeit. To avoid quarreling, the bookstore salesclerk decided to give her ten good $1 bills, taking back the counterfeit. This means that the bookstore salesclerk was out $3 (= cost of the book), plus the $10 bills he gave to the record-store salesclerk. Altogether, he lost $13. But only $10 were used in the whole transaction! Can you explain this?

67. Before they are blindfolded, three women are told that each one will have either a red or a blue cross painted on her forehead. When the blindfolds are removed, each woman is then supposed to raise her hand only if she sees a red cross and to drop her hand when she figures out the color of her own cross. Now, here's what actually happens. The three women are blindfolded and a red cross is drawn on each of their foreheads. The blindfolds are removed. After looking at each other, the three women raise their hands simultaneously. After a short time, one woman lowers her hand and says, "My cross is red." How did she figure it out?

68. Three women decide to go on a holiday to a Florida resort. They share a room at a hotel that is charging 1920s rates as a promotional gimmick. The women are charged only $10 each, or $30 in all. After going through his guest list, the manager discovers that he has made a mistake and has actually overcharged the three vacationers. The room the three are in costs only $25. So, he gives a bellhop $5 to return to them. The duplicitous bellhop knows that he cannot divide $5 into three equal amounts. Therefore, he pockets $2 for himself and returns only $1 to each woman.

Now, here's the conundrum. Each woman paid $10 originally and got back $1. So, in fact, each woman paid $9 for the room. The three of them together thus paid $9 × 3, or $27 in total. If we add this amount to the $2 that the bellhop dishonestly pocketed, we get a total of $29. Yet the women paid out $30 originally! Where is the other dollar?

Limits

69. Take the ratio of successive pairs of Fibonacci numbers. Express in limit notation what the ratio approaches:

$$\{1, 1, 2, 3, 5, 8, 13, 21, 34, 55, 89, 144, 233, 377, 610, 987, \ldots\}.$$

70. Draw any two equal intersecting lines, so that they bisect each other. Label the equal line segments r:

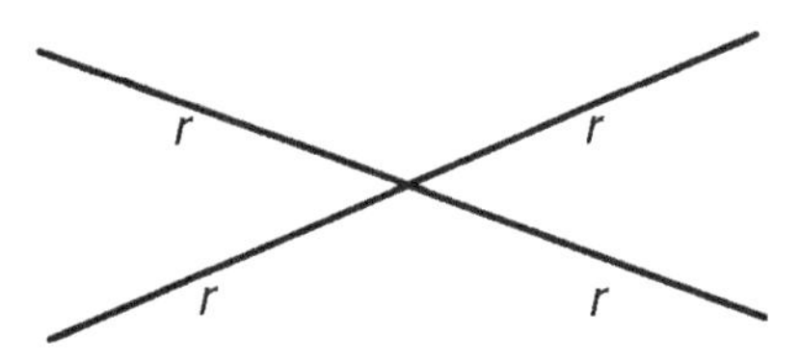

Keep on adding lines to the diagram that are of equal length to the original two, passing through the point of intersection. Each line drawn should itself be bisected into two parts of length r:

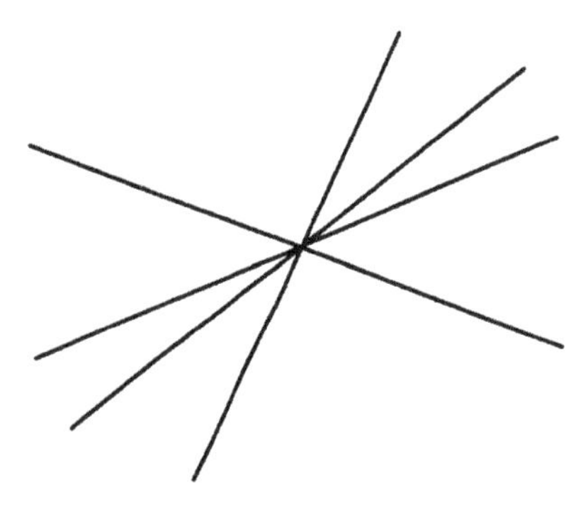

What figure does this technique, if continued indefinitely, approach? Can you prove it?

Further Reading

Barwise, Jon, and John Etchemendy. *The Liar*. Oxford: Oxford University Press, 1986.

Carroll, Lewis. *The Game of Logic*. New York: Dover, 1958.

Casti, John L., and Werner DePauli. *Gödel: A Life of Logic*. Cambridge, Mass.: Perseus, 2000.

Gardner, Martin. *Gotcha! Paradoxes to Puzzle and Delight*. San Francisco: Freeman, 1982.

Nagel, Ernest, and James R. Newman. *Gödel's Proof*. New York: New York University Press, 1958.

Rescher, Nicholas. *Paradoxes: Their Roots, Range, and Resolution*. Chicago and La Salle: Open Court, 2001.

Sainsbury, R. M. *Paradoxes*. Cambridge, U.K.: Cambridge University Press, 1995.

Salmon, W. C. (ed.). *Zeno's Paradoxes*. Indianapolis: Bobbs-Merrill, 1970.

Smullyan, Raymond. *The Riddle of Scheherazade and Other Amazing Puzzles, Ancient and Modern*. New York: Knopf, 1997.

———. *What Is the Name of This Book? The Riddle of Dracula and Other Logical Puzzle*s. Englewood Cliffs, N.J.: Prentice-Hall, 1978.

The Lo Shu Magic Square

Nobody before the Pythagoreans had thought that mathematical relations held the secret of the universe. Twenty-five centuries later, Europe is still blessed and cursed with their heritage.

ARTHUR KOESTLER (1905–1983)

ARRANGING THE FIRST NINE INTEGERS in a square pattern so that the sum of the numbers in each row, column, and diagonal is the same is called *Lo Shu* in China. This "magical" pattern was discovered four thousand years ago. The Chinese have always perceived it to have mystical properties. To this day, it is thought to provide protection against the evil eye when placed over the entrance to a dwelling or a room. Every fortune-teller uses it to cast fortunes. Amulets and talismans are commonly designed with Lo Shu inscribed in them.

Known appropriately as the magic square in English, Lo Shu spread from China to other parts of the world in the second century A.D. Around 1300, the Greek mathematician Emanuel Moschopoulos introduced it to Europe. Devising different kinds of magic squares became a veritable craze shortly thereafter. Like the Chinese, medieval astrologers perceived occult properties in them, using them to cast horoscopes. They also saw them as concealing coded cosmic messages. The eminent astrologer Cornelius Agrippa (1486–1535), for example, believed that a magic square of one cell (a square containing the single digit 1) represented the eternal perfection of God. Agrippa also took the fact that it is impossible to construct a 2×2 magic square to be proof of the imperfection of the four elements: air, earth, fire, and water.

But the significance of Lo Shu is not only found in the realm of the mystical. Magic squares have provided many insights into the nature of numbers and the design of mathematical techniques. The notion of algorithm stands out as particularly important in this regard. An algorithm is a technique that aims to "regularize" the solution of some specific problem or set of problems. Thus, given its importance to this area of mathematical method, Lo Shu, being the original magic square, surely belongs on the list of the ten greatest puzzles of all time.

The Puzzle

One version of the story of Lo Shu goes somewhat as follows. In ancient China, there was a huge flood. The people offered sacrifices to the god of the Lo River, to calm his anger. However, only one thing happened each time: a turtle appeared, crawling out of the river to walk nonchalantly around each sacrifice. The people saw the turtle as a sign from the river god who, they thought, kept rejecting their sacrifices. One day a child noticed a square on the shell of the turtle. In it were the first nine digits, arranged in three rows and columns. The child also noticed that the numbers along the rows, the columns, and the two diagonals consistently added up to 15. From this, the people realized the number of sacrifices that the river god required of them before he would be appeased.

Another version of the Lo Shu story has Emperor Yu the Great walking along the banks of the Lo River, when he saw a mysterious turtle crawl from the river. On its shell was a square arrangement of the first nine integers. Like the child, Yu noticed that the numbers in the square formed the pattern described previously, and he perceived the arrangement to constitute a message from the gods.

Whatever the truth of the matter, the original Lo Shu was a square made up of the first nine whole numbers, {1, 2, 3, 4, 5, 6, 7, 8, 9}, distributed in such a way that the numbers in the three rows, the three columns, and the two diagonals add up to 15. This is known as the **magic square constant**. Lo Shu is shown in this figure in its original form with figurate numbers (lines, dots, and circles):

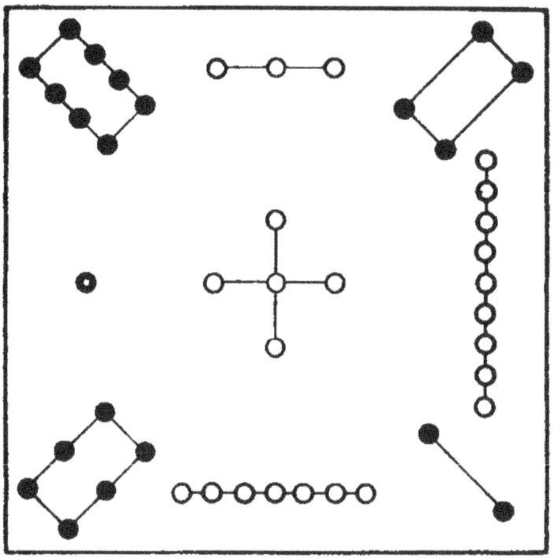

With decimal numbers, it looks like this:

8	3	4
1	5	9
6	7	2

Rows	**Columns**	**Diagonals**
8 + 3 + 4 = 15	8 + 1 + 6 = 15	8 + 5 + 2 = 15
1 + 5 + 9 = 15	3 + 5 + 7 = 15	4 + 5 + 6 = 15
6 + 7 + 2 = 15	4 + 9 + 2 = 15	

Lo Shu is known, more specifically, as a magic square of "order 3," a term indicating the number of cells in the square (3 × 3). A 4 × 4 square is called an "order 4" magic square, a 5 × 5 square an "order 5" magic square, and so on. In general, an $n \times n$ (= n^2) square is called an "order n" magic square.

The digits of Lo Shu can be arranged in several other ways to produce the magic square constant of 15. Here are two such arrangements:

4	9	2
3	5	7
8	1	6

8	1	6
3	5	7
4	9	2

Is there a method to the construction of magic squares? Or is it just a matter of trial and error? First and foremost, it would certainly be helpful if we had a general formula for determining what the magic square constant is. A magic square is made from a series of consecutive integers arranged into a square pattern. The last integer in the series is thus n^2, which is the order of the square. For example, Lo Shu consists of the consecutive integers from 1 to 9—{1, 2, 3, 4, 5, 6, 7, 8, 9}. The last digit is 9 or 3^2, which is the order of the square since it is called a 3 × 3 or "order 3" magic square. Similarly, in an "order 4" magic square, the last number is 4^2 (= 16); in an "order 5" magic square, it is 5^2 (= 25); and so on. In an "order n" magic square, therefore, the last number is n^2. With our summation technique (chapter 3), we can now set up an appropriate formula for the sum of the numbers in a magic square:

Sum of n numbers: $\frac{n(n+1)}{2}$

$\downarrow$

Sum of n^2 numbers in a magic square: $\frac{n^2(n^2+1)}{2}$

What we have done, in effect, is to replace n in the general formula with n^2:

$S_{(n)}$ (Sum of n numbers): $\frac{n\ (n+1)}{2}$

$\downarrow \quad \downarrow$

$S_{(n)}{}^2$ (Sum of n^2 numbers): $\frac{n^2(n^2+1)}{2}$

$n^2(n^2 + 1)$

When $(n^2 + 1)$ is multiplied by n^2, the result is $(n^4 + n^2)$: $n^2(n^2 + 1) = (n^4 + n^2)$.

Here's how the multiplication is carried out, step-by-step:

1. n^2 is multiplied by the first term in the expression $(n^2 + 1)$. This term is n^2. The result is: $n^2 \times n^2 = n^4$.
2. n^2 is then multiplied by the second term in the expression $(n^2 + 1)$. This term is 1. The result is: $n^2 \times 1 = n^2$.
3. The two results, n^4 and n^2, are added together: $(n^4 + n^2)$.

Let's simplify this formula (see the sidebar, if you have forgotten your high school math):

$$\frac{n^2(n^2+1)}{2} = \frac{(n^4+n^2)}{2}.$$

Now, let's apply this to Lo Shu, where $n = 3$:

$$\frac{(n^4+n^2)}{2} = \frac{(3^4+3^2)}{2} = \frac{90}{2} = 45.$$

This is the sum of the integers in a 3×3 square. If we divide this sum (45) by 3, we will get the magic square constant: $45 \div 3 = 15$. In general, we get the magic square constant by dividing the sum of the numbers in the square by n:

Sum of the numbers in a magic square: $\frac{n^2(n^2+1)}{2}$.

Dividing by n: we get: $\frac{n^2(n^2+1)}{2n}$.

This can be simplified as follows:

$$\frac{n^2(n^2+1)}{2n} = \frac{n \times \cancel{n}(n^2+1)}{2\cancel{n}} = \frac{n(n^2+1)}{2}.$$

We now have our general formula. Let's see if it works with Lo Shu. Substituting $n = 3$, we get:

$$\frac{n(n^2+1)}{2} = \frac{3(3^2+1)}{2} = \frac{3(10)}{2} = \frac{30}{2} = 15.$$

As can be seen, it does indeed generate the magic square constant. Is there anything else that can facilitate the construction process? Consider Lo Shu again. Note that it is an "odd order" square—it is a square constructed with an odd number of integers. All odd order squares have a middle cell. And the number that fills that cell can be determined by figuring out on how many rows, columns, and diagonals it occurs in the square. In the case of Lo Shu, it occurs on one row, one column, and the two diagonals (four in total).

Now, there are eight possible number triplets (made up with the first nine integers) that add up to 15. These are the rows, the columns, and the diagonals of the square:

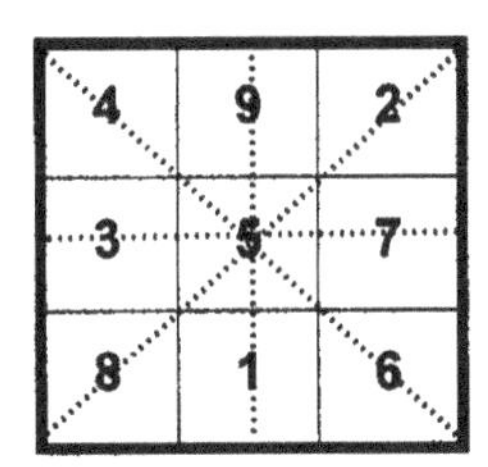

$9 + 5 + 1 = 15$

$9 + 4 + 2 = 15$

$8 + 6 + 1 = 15$

$8 + 5 + 2 = 15$

$8 + 4 + 3 = 15$

$7 + 6 + 2 = 15$

$7 + 5 + 3 = 15$

$6 + 5 + 4 = 15$

We established that the middle number appears in four such triplets. The only one that appears four times in the previous list is 5:

$9 + \underline{5} + 1 = 15$

$8 + \underline{5} + 2 = 15$

$7 + \underline{5} + 3 = 15$

$6 + \underline{5} + 4 = 15$

In this way, we have identified the middle number. A similar line of reasoning can be applied to magic squares of increasing odd order.

Mathematical Annotations

Although the square is the oldest and most common one, magic figures of different shapes have been devised. Magic cubes, for instance, are constructed with the numbers arranged in cubical form so that each row of numbers running parallel with any of the edges, and also with any of the four great diagonals, will have the same magic constant. In the following magic cube, that constant is 42, as readers can confirm for themselves:

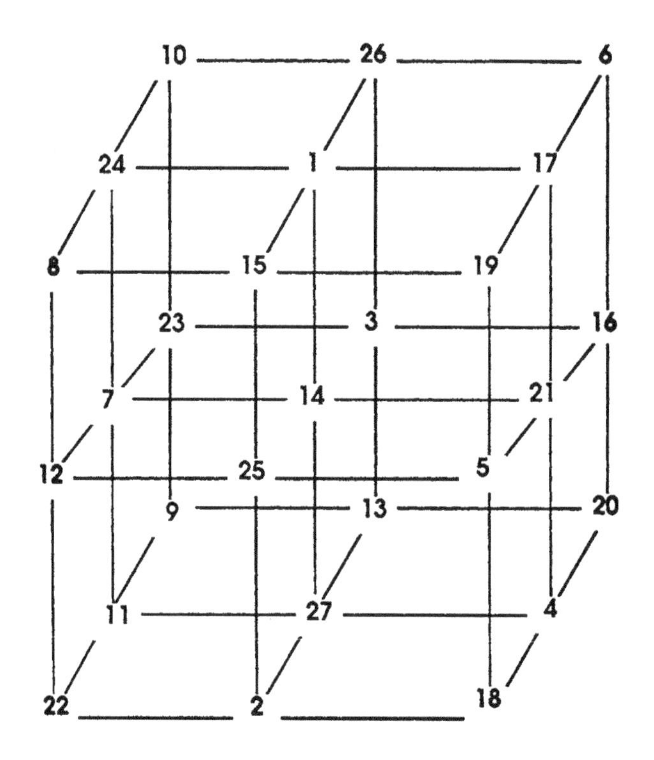

Magic figures put the spotlight on numerical pattern, in and of itself. As such, they are exercises in "pure" mathematical thinking.

ALBRECHT DÜRER (1471–1528)

Dürer was born in Nuremberg, Germany. His art and his writings had a profound influence on sixteenth-century artists. He is especially famous for having incorporated the use of tones of varying darkness in his paintings to give the illusion of three-dimensional form. For this reason, he is considered one of the founders of perspective painting.

"Magical" Number Patterns

In order to see what magic squares hold in store, let's look at one of the most famous of all time—the Dürer magic square—a square that has caught the attention of so many mathematicians that it would take several pages just to name them. It is named after its constructor, the great German Reformation artist Albrecht Dürer. Dürer put the square in his famous 1514 engraving *Melancholia*. Nearly two centuries after that, the Swiss mathematician Leonhard Euler (chapter 4) became so mesmerized by it that he constructed forty-eight versions of the square himself. Dürer's square is an "order 4" square, consisting of the first sixteen numbers. Its magic square constant is 34.

The magic square constant formula (given earlier):

$$\frac{n\,(n^2+1)}{2}$$

$n = 4$.

Therefore:

$$\frac{n\,(n^2+1)}{2} = \frac{4\,(4^2+1)}{2} = \frac{4\,(16+1)}{2} = \frac{4\,(17)}{2} = 34.$$

16	3	2	13
5	10	11	8
9	6	7	12
4	15	14	1

The square has many "magical" properties. For example, in addition to appearing in each row, column, and diagonal, the magic square constant of 34 also appears as follows:

- in the sum of the digits in the four corners (16 + 13 + 4 + 1 = 34)
- in the sum of the four digits in the center (10 + 11 + 6 + 7 = 34)
- in the sum of the digits 15 and 14 in the bottom row and the digits 3 and 2 facing them in the top row (15 + 14 + 3 + 2 = 34)
- in the sum of the digits 12 and 8 in the right-hand column and of 9 and 5 facing them in the left-hand column (12 + 8 + 9 + 5 = 34)
- in the sum of the digits of each of the four squares in the corners (16 + 3 + 5 + 10 = 34; 2 + 13 + 11 + 8 = 34; 9 + 6 + 4 + 15 = 34; 7 + 12 + 14 + 1 = 34)

The square contains many other interesting patterns. It would take a treatise of its own to discuss them. Dürer's square is not the first 4th-order square to have been devised. Archaeologists have found one in an inscription at Khajuraho, India, from the twelfth century:

7	12	1	14
2	13	8	11
16	3	10	5
9	6	15	4

It is known as a "diabolic" square because it retains its magical properties even if the bottom row is shifted from one side to another. Incidentally, there are a mind-boggling 880 ways to construct a 4th-order magic square.

A magic square of order 5 (which has a magic square constant of 65) has many more possible arrangements—275,305,224 in all! Here is one of them:

17	24	1	8	15
23	5	7	14	16
4	6	13	20	22
10	12	19	21	3
11	18	25	2	9

Perhaps the most extraordinary of all magic squares was the order 8 magic square devised by Benjamin Franklin (1706–1790), the great American public official, writer, scientist, and printer:

52	61	4	13	20	29	36	45
14	3	62	51	46	35	30	19
53	60	5	12	21	28	37	44
11	6	59	54	43	38	27	22
55	58	7	10	23	26	39	42
9	8	57	56	41	40	25	24
50	63	2	15	18	31	34	47
16	1	64	49	48	33	32	17

Constructed with the first sixty-four integers, Franklin's square contains a host of astonishing numerical oddities, such as the following:

- Its magic square constant is 260; and exactly half this number, 130, is the magic square constant of each of the four 4×4 squares that are quadrants of the larger square.
- The sum of any four numbers equidistant from the center is also 130.
- The sum of the numbers in the four corners plus the sum of the four center numbers is 260.
- The sum of the four numbers forming any little 2×2 square within the main square is 130.
- There are many more.

It is truly perplexing to contemplate how Franklin ever could have devised this masterpiece. Incidentally, Leonhard Euler came up with his own truly magical version of an order 8 square. The four quadrant squares within it have the magic square constant of 130, as in Franklin's square. But the unique property of Euler's square is that if you take the knight chess piece—which moves on the chessboard in an L-pattern—and start at 1 in the top left corner of the square, you will land on every number, from 1 to 64, once and only once!

1	48	31	50	33	16	63	18
30	51	46	3	62	19	14	35
47	2	49	32	15	34	17	64
52	29	4	45	20	61	36	13
5	44	25	56	9	40	21	60
28	53	8	41	24	57	12	37
43	6	55	26	39	10	59	22
54	27	42	7	58	23	38	11

Algorithms

The study of magic squares has had a significant impact on the development of the concept of algorithm. This is defined as a step-by-step method for solving specific kinds of problems. Constructing magic squares is largely a matter of trial and error. However, for some cases, an algorithm can be derived. At the very least, it can be tried out to see whether it generates a magic square or not.

ALGORITHMS

An **algorithm** is a systematic technique used for solving a problem that involves a sequence of steps.

Here's an algorithm for putting on shoes and socks:

1. Put on the socks in any order and on either foot.
2. Put on the two shoes in symmetrical fashion: the left shoe on the left foot, the right shoe on the right foot.
3. Steps 1 and 2 cannot be reversed.

Here is an algorithm for a 4th-order square. First, we draw intersecting lines through the diagonals:

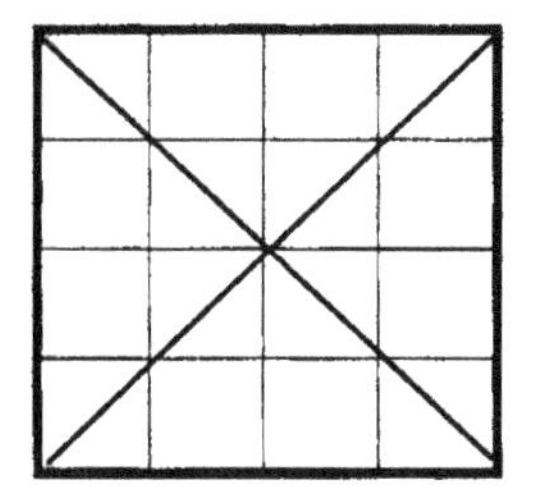

Next, put in the numbers as if they were consecutive, leaving blank those that are crossed out by the intersecting lines. Start with 1 in the upper-left corner cell. Since it is crossed, leave it blank. Pass on to the next one to the right. Since it is empty, put the next number in it, 2. The third cell is also empty, so put 3 in it. The fourth cell is crossed, so leave it empty. Proceed in this fashion until you reach the last cell in the bottom right-hand corner.

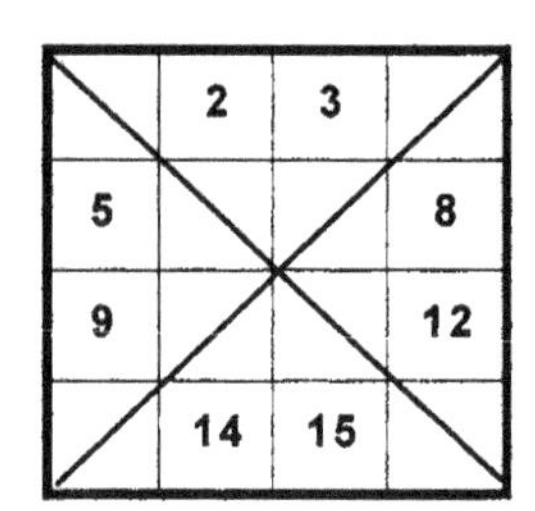

Now, begin at the lower-right corner, and move across the rows leftward, recording only the numbers in the cells cut by the diagonal lines. So, start by putting 1 in the right-hand corner. The next two are filled. When you reach the bottom-left corner, put in the next number, which is 4, since 2 and 3 have already been used. And so on.

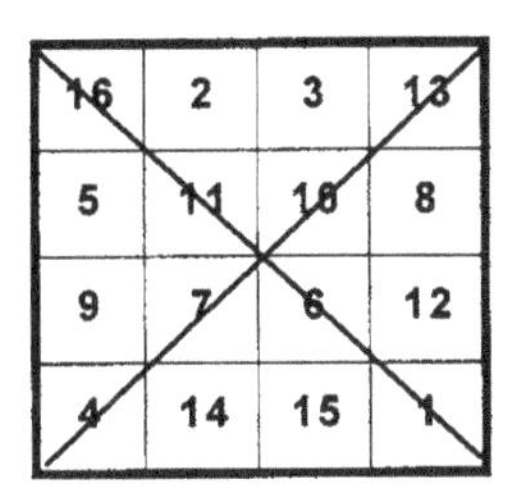

This same algorithm can be used to generate an order 8 magic square. Constructing such a square is left as an exercise in the Explorations section.

Another algorithm for constructing an odd order square (a square of order 3, 5, 7, etc.) is attributed to the mathematician Simon de la Loubère (1642–1729) in 1693, although he probably learned about it during his travels to Asia. Let's use his algorithm on an order 5 magic square—a square consisting of the first twenty-five numbers, with a magic square constant of 65:

1. Place 1 in the central upper cell:

		1		

2. Proceed diagonally upward to the right and place the next digit, 2, in an imaginary square outside the actual square. Because the 2 is outside the square, bring it to the bottom of the column in alignment with it:

			2		
		1	↓		
			↓		
			↓		
			↓		
			2		

3. Put the next digit, 3, diagonally upward to the right of 2:

			2		
		1	↓		
			↓		
			↓		
			↓	3	
			2		

4. Using the same upward right diagonal movement, insert the 4 in the imaginary cell to the right of 3 and, subsequently, at the opposite end of the row.

			2		
		1			
4	←	←	←	←	4
				3	
			2		

5. Insert 5 diagonally upward to the right of 4:

			2		
		1			
	5				
4					4
				3	
			2		

6. The same movement pattern cannot be followed to insert the 6 because the cell that is diagonally upward to the right of 5 is already occupied. The 6 is therefore written below the 5.

			2		
		1			
	5				
4	6				4
				3	
			2		

7. Proceed in this fashion to complete the square (readers are invited to do so by themselves):

	18	25	2	9	
17	24	1	8	15	17
23	5	7	14	16	23
4	6	13	20	22	4
10	12	19	21	3	10
11	18	25	2	9	

Incidentally, we can start by putting the 1 in any cell. However, this will generate a square that is magic in the rows and the columns only—not in the diagonals.

Reflections

Even though they may have no practical applications, magic squares are nevertheless interesting in themselves, because they impel us to think about number patterns in "pure" terms. And who knows, some day we may find magic squares cropping up in nature and human affairs, as the Fibonacci numbers do.

Magic squares provide a clue about why the early histories of mathematics and magic overlap considerably. In their origins, both sought to do the same thing—unravel hidden patterns. In antiquity, no distinction was made between *numeration* and *numerology* (the science that studied the purported divinatory properties of numbers). Numerology started with the Pythagoreans, who taught that numbers were the language of the cosmos (as discussed previously). The ancient Israelites held a similar belief, establishing the art of **gematria** on the view that the letters of any word or name could be interpreted as digits and rearranged to form a number that contained a secret message. The earliest recorded use of gematria, actually, was in the eighth century B.C. by King Sargon II of Babylon, who built the wall of the city of Khorsbad exactly 16,283 cubits long because this was the numerological value of his name.

As it turns out, the Lo Shu square also has hidden properties that impart a true numerological aura to it. For instance, if the successive terms of the Fibonacci series starting at 3 and ending with 144—{3, 5, 8, 13, 21, 34, 55, 89, 144}—are matched in order with the integers in the Lo Shu square, a new square is formed, as shown:

8	1	6
3	5	7
4	9	2

→

89	3	34
8	21	55
13	144	5

Original Magic Square Number	Replaced with Fibonacci Number
1	3
2	5
3	8
4	13
5	21
6	34
7	55
8	89
9	144

The new square has the following property: the sum of the products of the three rows equals the sum of the products of the three columns:

Row Products			Column Products		
$89 \times 3 \times 34$	=	9,078	$89 \times 8 \times 13$	=	9,256
$8 \times 21 \times 55$	=	9,240	$3 \times 21 \times 144$	=	9,072
$13 \times 144 \times 5$	=	9,360	$34 \times 55 \times 5$	=	9,350
Sum:		27,678	Sum:		27,678

Whatever the significance of this amazing result, it imbues Lo Shu with even more mystical power than it already is perceived to have. Incidentally, only after the Renaissance was numerology relegated to the status of a pseudoscience. Paradoxically, the Renaissance at first encouraged interest in the ancient occult art and its relation to mathematical inquiry. The Roman Catholic Church and the new Protestantism, however, turned sharply against it in the fifteenth and sixteenth centuries. As a result, mathematics was no longer enshrouded in mystical symbolism, as it had been in the ancient world.

But the connection between magic, symbolism, and mathematics has hardly been severed. Mathematical patterns continue to cast a "magical spell" over us. A thick volume could be written about the many meanings ascribed to specific numbers across the world and across history. People tend to think of certain things, such as dates, street addresses, or certain specific numbers, as having great significance. Human beings seem to possess the basic Pythagorean notion that the world itself is a magical pattern of small numbers arranged in patterns.

Explorations

Magic Squares

71. Can you arrange the first nine even numbers, {2, 4, 6, 8, 10, 12, 14, 16, 18} into an order 3 magic square? What is its magic square constant?

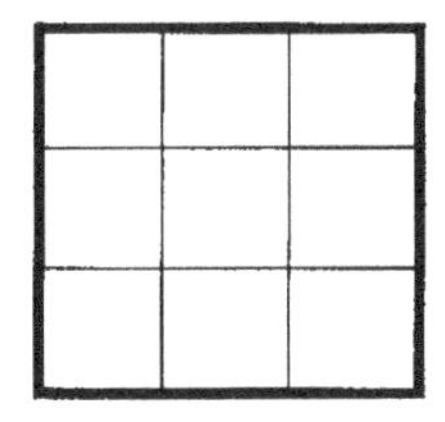

72. Can you arrange the following nine consecutive numbers {4, 5, 6, 7, 8, 9, 10, 11, 12} into an order 3 magic square? What is its magic square constant?

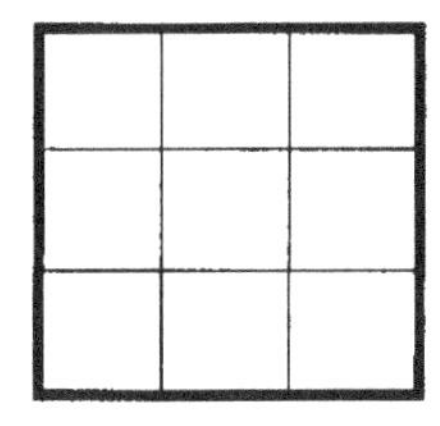

73. Can you arrange the following nine consecutive decimals {0.25, 0.50, 0.75, 1.00, 1.25, 1.50, 1.75, 2.00, 2.25} into an order 3 magic square? The magic square constant is 3.75. The central cell is 1.25.

	1.25	

74. The following tough nut is due to none other than the great British puzzlist Henry E. Dudeney. Can you arrange the following nine prime numbers {1, 7, 13, 31, 37, 43, 61, 67, 73} into an order 3 magic square? The magic square constant is 111.

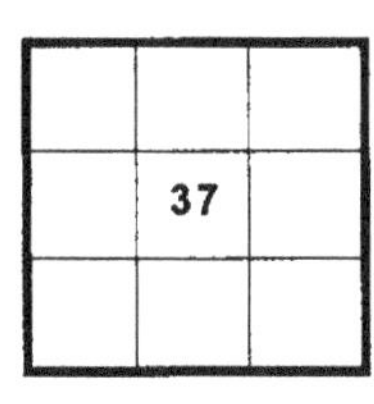

75. The following puzzle is due to another great puzzlist, Lewis Carroll, who used the postal values of his day to challenge puzzle enthusiasts. In Victorian times, postage values were expressed in half units. Can you arrange the following postage stamps {1d, $1\frac{1}{2}$d, 2d, $2\frac{1}{2}$d, 3d, $3\frac{1}{2}$d, 4d, $4\frac{1}{2}$d, 5d} into an order 3 magic square? What is its magic square constant?

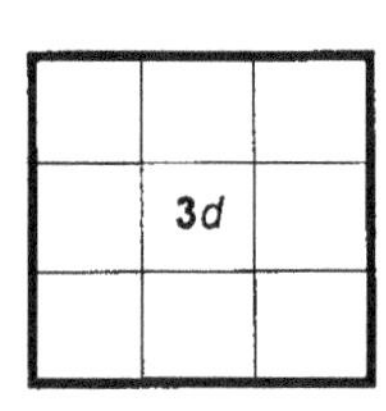

76. Now, try your hand at constructing a 4th-order magic square, with a magic square constant of 102. This is a difficult puzzle. Ignore it if you get too bogged down. To help you, some of the cells have been filled in. Moreover, note that the smallest number is 1 and the largest is 71 and that all the numbers (except 1) are prime.

	71		23
53	11		
29			47

Algorithms

77. Construct an 8th-order square, using the algorithm described in this chapter for an order 4 square.

78. Can you derive an algorithm for an order 3 magic square, using the ideas discussed in this chapter?

Further Reading

Andrews, W. S. *Magic Squares and Cubes*. New York: Dover, 1960.

Benson, W. H., and O. Jacoby. *Magic Cubes: New Recreations*. New York: Dover, 1981.

———. *New Recreations with Magic Squares*. New York: Dover, 1976.

Clawson, Calvin C. *Mathematical Mysteries: The Beauty and Magic of Numbers*. Cambridge, Mass.: Perseus, 1996.

Gardner, Martin. *Mathematical Magic Show*. Washington, D.C.: Mathematical Association of America, 1990.

———. Martin. *Mathematics, Magic, and Mystery*. New York: Dover, 1956.

Heath, Royal V. *Mathemagic: Magic, Puzzles, and Games with Numbers*. New York: Dover, 1953.

Joseph, George G. *The Crest of the Peacock: Non-European Roots of Mathematics*. Harmondsworth, U.K.: Penguin, 1991.

Kasner, Edward, and John Newman. *Mathematics and the Imagination*. New York: Simon and Schuster, 1940.

Li, Yen, and Du Shiran. *Chinese Mathematics: A Concise History*. New York: Oxford University Press, 1987.

Pickover, Clifford A. *The Zen of Magic Squares, Circles, and Stars*. Princeton, N.J.: Princeton University Press, 2002.

Simon, William. *Mathematical Magic*. New York: Dover, 1964.

Stewart, Ian. *The Magical Maze: Seeing the World through Mathematical Eyes*. New York: John Wiley & Sons, 1997.

The Cretan Labyrinth

In these years we are witnessing the gigantic spectacle of innumerable human lives wandering about lost in their own labyrinths, through not having anything to which to give themselves.

JOSÉ ORTEGA Y GASSET (1883–1955)

IF WE WERE TO GO INSIDE THE BURIAL chambers of the ancient pyramids, such as the one at Giza in Egypt, we could not help but be overwhelmed by the intricate system of intertwining passages within them. Undoubtedly, the architecture of such tombs was designed to challenge a dead person's soul to find "the one true path" to the afterlife. Buildings designed in such a way were called **labyrinths**. While the mysticism may have faded, to this day the labyrinth concept (also known as *maze* in English) nevertheless continues to be used to provide challenges of various sorts. Psychologists, for instance, use mazes to evaluate problem-solving skills in animals and humans alike. And toy mazes are among the most popular types of games given to children today, mainly because they are thought to sharpen logical skills, while at the same time providing recreation.

The first known labyrinth was a prison built on the island of Crete. According to legend, the prison was constructed by the Athenian craftsman Daedalus for King Minos of Crete as an "architectural puzzle." Minos had the dungeon built to avenge the death of his son Androgeus at the hands of a group of unknown Athenians. Adding to his woe was the fact that his wife, Pasiphae, had fallen in love with a bull and given birth to a half-human, half-bull beast called the Minotaur (literally, "the bull of Minos"). Shamed by this event and aching to exact his revenge on Athens, Minos sent seven young Athenian men and women every year into the prison. At its center he put the voracious Minotaur, who was eager to destroy anyone who ventured there. Theseus, the son of King Aegeus of Athens, offered to go as one of those to be sacrificed. Ironically, Minos's clever daughter, Ariadne, had fallen in love with Theseus. So, she gave her beloved a sword with which to kill the Minotaur and a thread to mark his path through the labyrinth. Theseus slew the Minotaur and emerged to reunite with Ariadne, finding his way back simply by following the path marked by the thread. Aegeus had instructed Theseus to raise a white sail on his ship after he had accomplished his mission. But Theseus forgot to do so, and, as legend has it, when his father saw the ship returning with black sails, he threw himself into the sea, which was thereafter called the Aegean. Archaeologists have discovered a palace located in the Cretan city of Knossos that may have been the site of the mythical labyrinth, because it has many passageways like those described in the legendary account of Minos's prison.

Now, what possible connection to mathematics does the Cretan Labyrinth have, the reader may ask? The labyrinth is essentially a puzzle in topology. As such, it is an overarching idea pattern that mathematicians have used throughout the ages to study the nature of various topological structures. For this reason, the very first labyrinth of history belongs on the list of the top ten puzzles of all time.

The Puzzle

No one really knows what the Cretan Labyrinth actually looked like. Its most likely shape is found on ancient coins discovered at Knossos, the probable site of the Cretan Labyrinth:

The solution to the Cretan Labyrinth is straightforward. By entering at the opening and following its single winding path, you will reach the center. The Cretan Labyrinth is called a *unicursal* Eulerian graph (chapter 4)—a graph with one path through it. Mazes with alternative paths pose a much greater challenge, because there is no algorithm for solving them. However, some useful suggestions have been put forward by mathematicians over the years. The following are paraphrased from Edouard Lucas (whom we encountered in chapter 6):

- As you go through the maze, constantly keep looking ahead along a path to see if it ends up being a "dead end"; if so, avoid it and take another one at some juncture.
- Whenever you come to a new juncture, look ahead to scrutinize the path as open or dead.
- If on a path you come to an old juncture or a dead end, turn and go back the way you came.
- Never enter a path marked on both sides.

Consider the following maze. The objective is to start at the opening at the bottom and find a path to the area marked by the large dot. As readers can confirm for themselves, the previous guidelines are indeed useful. The path shown on the following page is the solution.

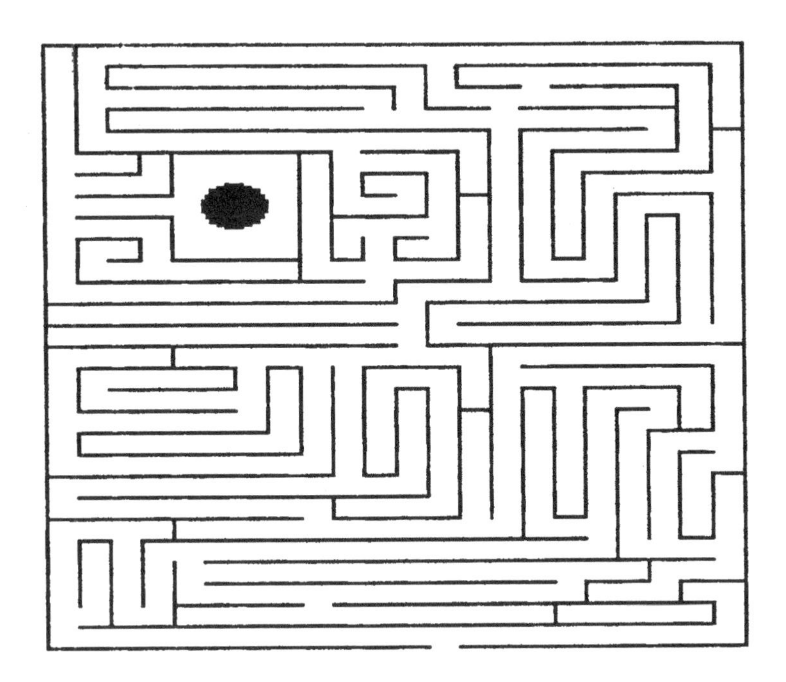

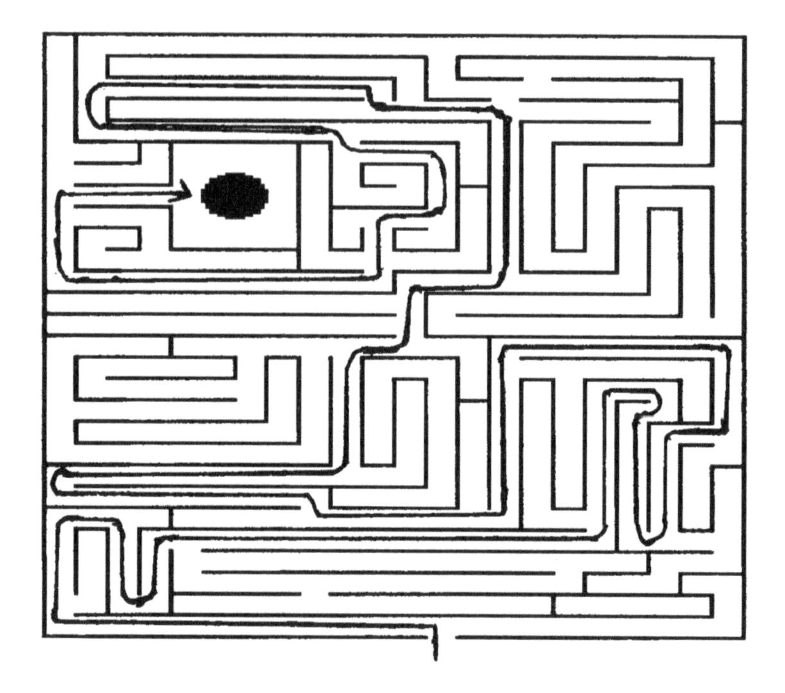

The Cretan Labyrinth has appealed to rulers, philosophers, mathematicians, artists, and writers alike. The later Roman emperors had copies of it embroidered on their robes. It has been found etched on the walls of many early Christian churches.

The labyrinth concept is universal. One of the oldest labyrinthine designs is found carved in the stone wall of a five-thousand-year-old grave in Sicily. Similar carvings have been discovered throughout the world. Labyrinths have been used throughout history and across cultures to ward off evil, invoke supernatural powers, and test the intelligence of heroes. As mentioned, the Egyptians designed their pyramids as labyrinths. They also constructed some of their buildings as labyrinths. The largest one was the Great Labyrinth, a huge building constructed around 2000 B.C. in northern Egypt, with three thousand rooms. The ancient city of Troy was also designed with labyrinthine paths, providing protection against invaders by confusing them. In Java, Sumatra, and India, the labyrinth design has been used from time immemorial as a symbol of inner peace. The Navajo people in the United States have always considered the labyrinth to be a representation of how the world was created. The floors of many medieval churches had labyrinthine designs in them, to symbolize the tortuous journey of individuals toward salvation. One of the largest can be found at Chartres in France. From the Renaissance onward, many European gardens were designed as mazes walled by clipped hedges. Two of the best known were built in the seventeenth century—at Hampton Court in London and in the

Palace of Versailles, which is adorned with thirty-nine fountains and various statues depicting characters from Aesop's fables.

Mathematical Annotations

The concept of the labyrinth has been used to study the structure of graphs, given that the idea behind the construction of a labyrinth is to identify the optimal path in a network. Ultimately, all geometric figures are graphs and can be analyzed as such. One of the most important developments related to the concept of graphs is coordinate geometry, as will be discussed briefly in the next section.

Coordinate Geometry

A slew of problems in geometry require finding an optimal path. Here is a typical one:

> The length of a small rectangular floor is twice its width. The area of the floor is 32 square feet. A bug in the lower left corner wants to get to the opposite corner. What is the shortest path for the bug? What is the length of that path?

First, draw the rectangular floor. On it, let x stand for its width and $2x$ for its length. The latter expression simply indicates that the length is twice the width.

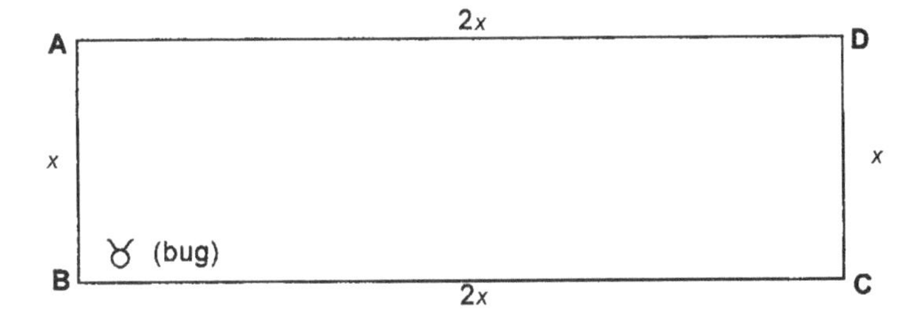

The optimal path for the bug is the diagonal path to the opposite corner:

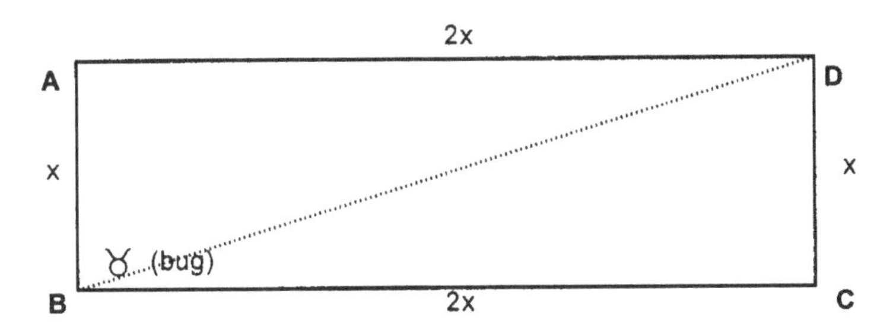

The diagonal is the hypotenuse of the right-angled triangle **DBC**, with sides of two lengths, x and $2x$ feet. So, if we can determine what these lengths are, we could then use the Pythagorean theorem (chapter 5) to establish the length of the diagonal. How do we do this?

We are told that the area of the floor is 32 square feet. From school geometry, recall that the area of a rectangle is the product of the length times the width. In this case, the length is $2x$ and the width is x. Multiplying these together, we get

Area of floor:
$(2x)\,(x) = 32$
$2x^2 \quad = 32$

Dividing both sides of the equation by 2:
$x^2 = 16$

Taking the square root:
$x = 4$

We now know that the width is 4 feet. Since the length is twice this, it is 8 feet. These are the lengths of the sides of our right-angled triangle **DBC**:

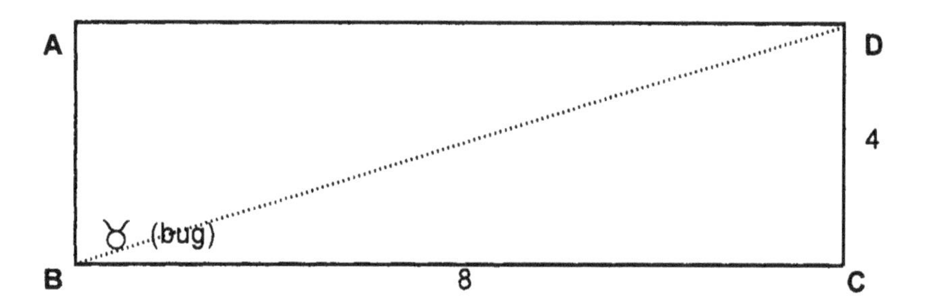

We can now use the Pythagorean theorem to determine the length of its hypotenuse, **DB**:

$\mathbf{DB}^2 = 4^2 + 8^2$

$\mathbf{DB}^2 = 16 + 64$

$\mathbf{DB}^2 = 80$

$\mathbf{DB} = \sqrt{80} = 8.94$

Thus, the optimal path for our bug is 8.94 feet long. A more challenging version of this problem is called the Spider and the Fly Puzzle. It is assigned as an exercise in the Explorations section.

The study of optimal paths brings out the fact that arithmetic, algebra, and geometry are interrelated. This was known to the ancient mathematicians. However, the formal amalgamation of these disciplines had to await the work of the French mathematician and philosopher René Descartes (1596–1650). He called his amalgamation **analytic geometry** (the branch of mathematics that studies geometric figures and properties by converting them into algebraic form). The basic notion in analytic geometry is that of intersecting "number lines." A number line, in fact, is itself a rudimentary geometric representation that shows the continuity between positive and negative numbers and a one-to-one correspondence between a specific number and a specific "point" on the line:

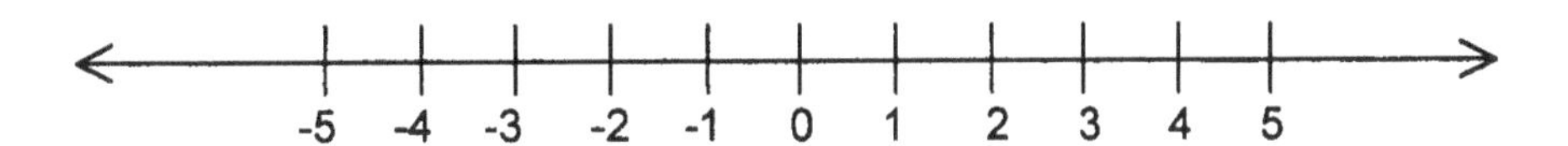

Descartes simply drew two number lines intersecting at right angles. He called the horizontal line the "*x*-axis," the vertical one the "*y*-axis," and their point of intersection the "origin." This system of two perpendicular intersecting number lines is now called the **Cartesian plane**, in honor of Descartes.

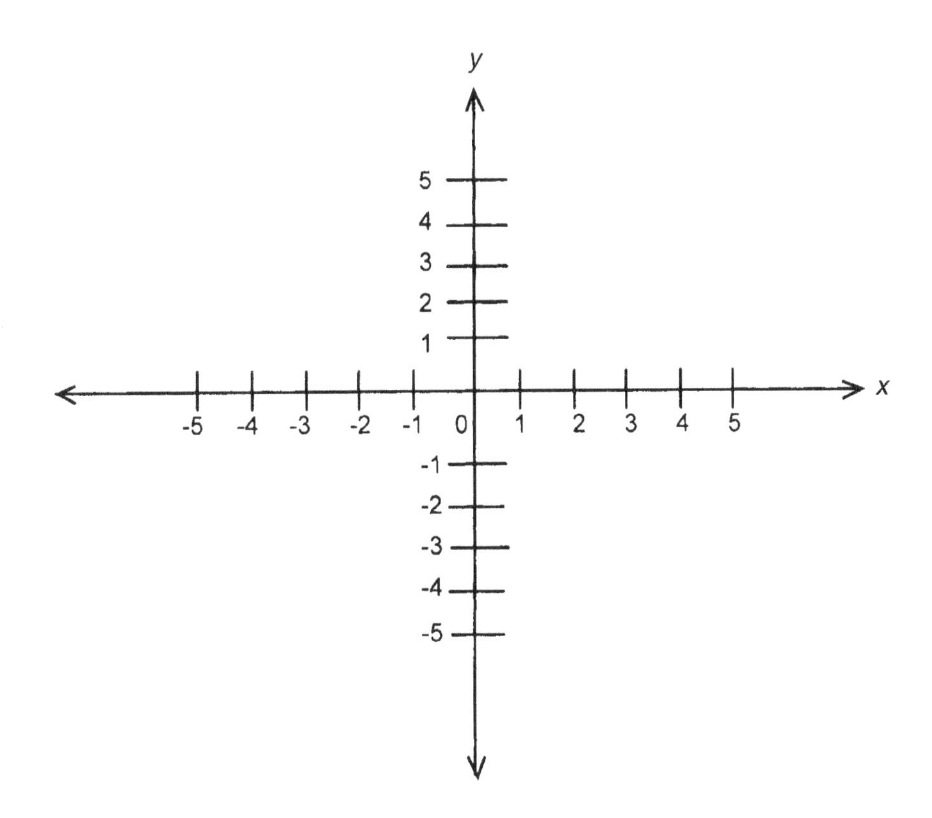

It is also called a **coordinate system**, because the plane can now be conceived as a system of points that are determined by their positions in relation to the two axes, called "coordinates." For example, the paired coordinates for point A in the following figure are (2,1). This means that point A is two units to the right of the y-axis, and one unit directly above the x-axis. In addition, the figure shows several other points—B, C, and D—and their coordinates:

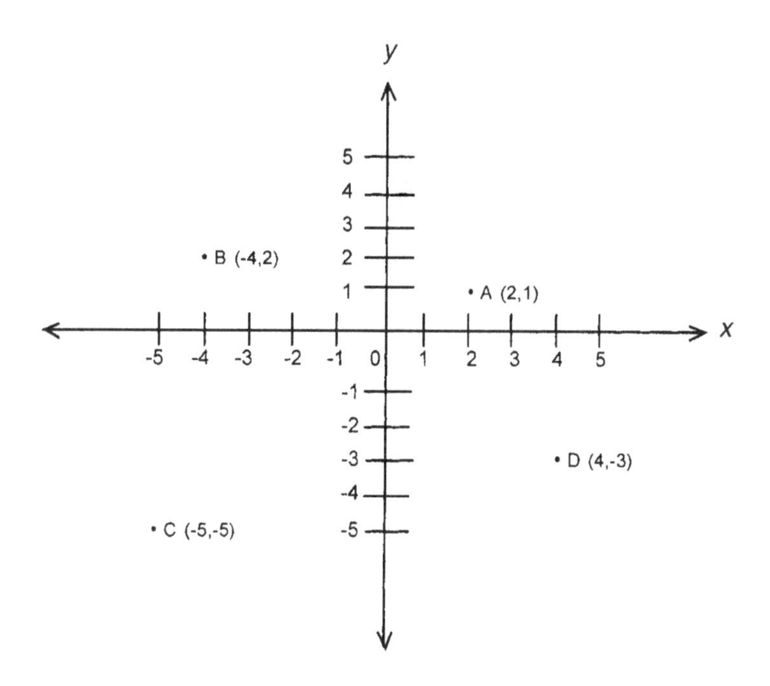

With this system of assigning coordinates, an equation such as $2x + y = 2$ can be plotted to reveal its underlying "geometric form," which turns out to be a line. Note that we can plot the points that the line goes through by determining the solutions to the equation in coordinate terms (x,y). Some of the solutions are (–2, 6), (–1, 4), (0, 2), (1, 0), and (2, –2). These are obtained as follows:

$$2x + y = 2.$$

Subtracting $2x$ from both sides:

$$y = 2 - 2x.$$

Therefore:

If $x = -2$ then $y = 6$

If $x = -1$ then $y = 4$

If $x = 0$ then $y = 2$

If $x = 1$ then $y = 0$

If $x = 2$ then $y = -2$

If you plot these points on a coordinate system and then connect them with a smooth line, you will see that they lie on a straight line:

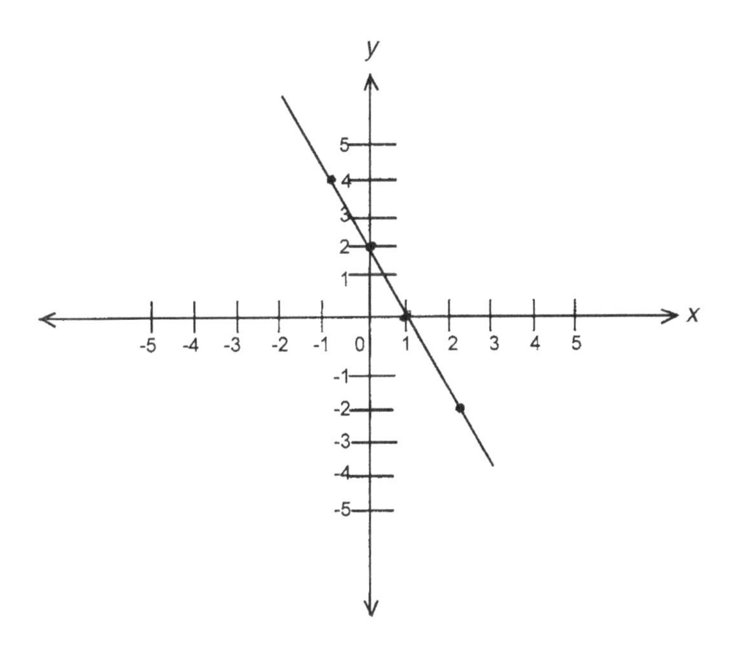

Any point that lies on the line has coordinates that satisfy the equation $2x + y = 2$, and any pair of numbers (x,y) that satisfies the equation will be a point on the line. Clearly, analytic geometry allows us to relate a type of equation to a type of geometrical figure. The equation $x^2 + y^2 = 25$, for instance, turns out to be the equation of a circle, as readers can confirm for themselves by assigning values to x and y and then plotting them on graph paper:

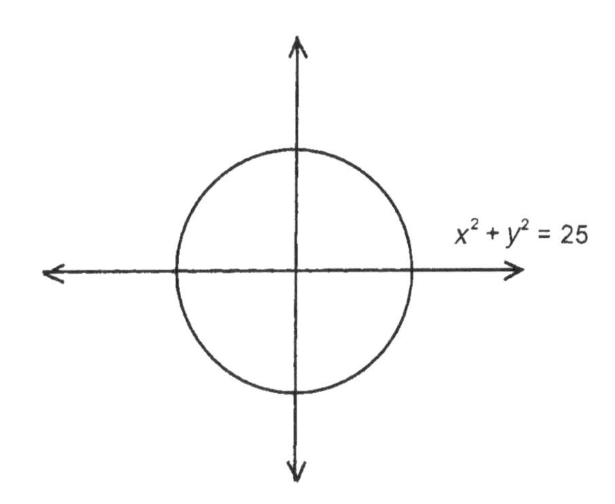

Analytic geometry has become the basis for making maps, for analyzing functions of all kinds, for developing theorems, for determining optimal paths—and the list could go on and on.

The Pythagoreans

The discussion of mazes, optimal paths, and analytic geometry takes us—in a labyrinthine fashion, no less—to the core of what mathematics is all about: the study of pattern. The Pythagoreans founded mathematics as the science of pattern. They were a truly extraordinary group. During a time when women were largely excluded from mathematics and philosophy, the Pythagoreans welcomed women as equals, providing them with a rare opportunity to participate in the fields of philosophy and mathematics. Pythagoras's wife, Theano, became an accomplished cosmologist and healer. She and her daughters, although persecuted, spread the Pythagorean philosophy throughout ancient Greece and Egypt.

The Pythagoreans claimed that mathematics was the language by which the world could be interpreted. They also noticed the relation between numbers and geometric figures, long before Descartes did. For example, they defined **triangular numbers** as those that showed a triangular pattern and **square numbers** as those that showed a square pattern. The numbers 1, 3, 6, and 10 are triangular, and 1, 4, 9, and 16 are square, because they can be represented as follows:

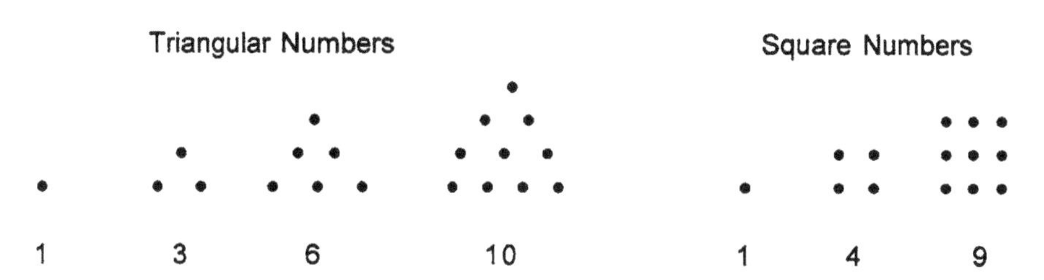

This Pythagorean insight was extended by later Greek mathematicians to argue that all numbers had analogues in the geometric domain. For example, they showed that the sum of two numbers $(a + b)$ corresponded to the addition of line segments and subtraction $(a - b)$ to the subtraction of line segments:

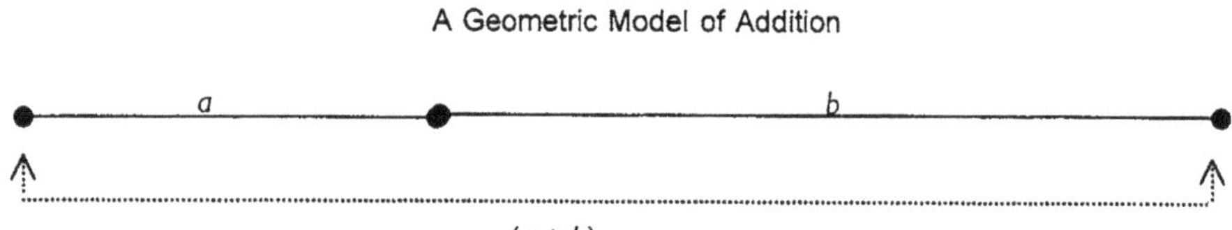

A Geometric Model of Subtraction

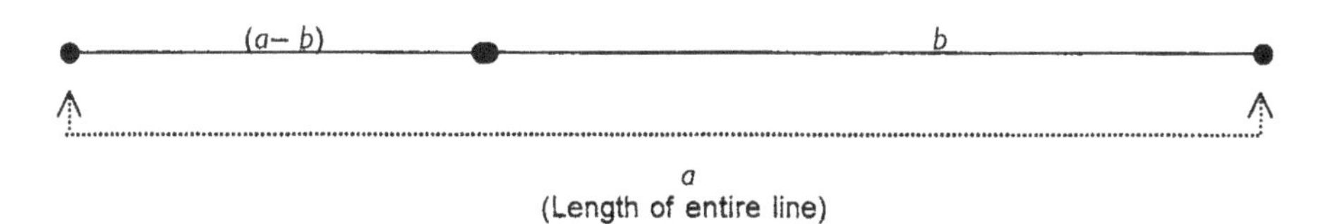

The first figure shows that addition corresponds to the joining up of two line segments, a and b, to produce the line $(a + b)$. The second figure shows that if a line with the length a is segmented into two parts, one of which has the length b, the remaining part will have the length $(a - b)$—which represents the length left over when b is removed from a.

Of all the patterns discovered by the Pythagoreans that show a relation between numbers and geometric figures, none is probably more important than the so-called **Pythagorean triples**—sets of three numbers, $\{a, b, c\}$, for which the relation $c^2 = a^2 + b^2$ is true. This relation reflects, of course, the fact that the square of the hypotenuse of a right triangle (c) is equal to the sum of the squares of the other two sides (a, b).

For the sake of historical accuracy, it should be mentioned that this relation was known far and wide before it was proved by Pythagoras. And virtually all the ancient builders probably had a practical knowledge of it. Clay tablets dating back to nearly 2000 B.C., for instance, reveal that the ancient Babylonians knotted ropes to make "3-4-5" right triangles—because $5^2 = 3^2 + 4^2$. They clearly had a practical knowledge of the Pythagorean theorem, and they were also familiar with many Pythagorean triples—3, 4, 5 ($3^2 + 4^2 = 5^2$); 6, 8, 10 ($6^2 + 8^2 = 10^2$); 5, 12, 13 ($5^2 + 12^2 = 13^2$); 8, 15, 17 ($8^2 + 15^2 = 17^2$); and so on.

Reflections

As a final reflection, I would like to emphasize one last time that puzzles are not only enjoyable in themselves, they also have the ability to illustrate basic notions of mathematics. I hope that the reader will come away from this book with a new perspective on puzzles and their relation to mathematical discovery. If I have done nothing more than to show this, writing the book will have been worthwhile.

Explorations

Labyrinths

79. Here is a much more difficult version of the Cretan Labyrinth. Or is it? Find out.

80. Can you find a path through the following maze?

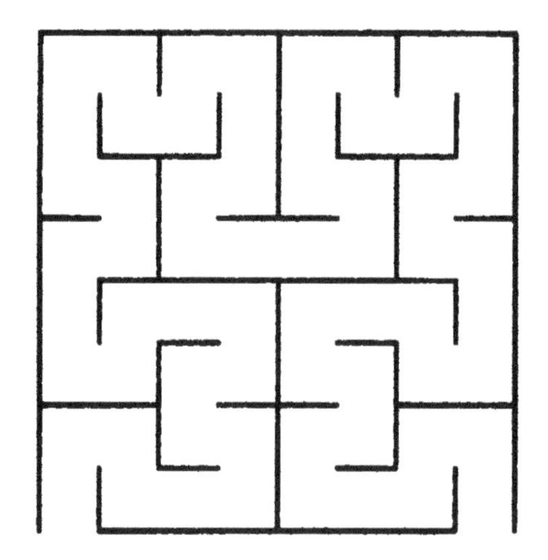

81. Here is a more difficult version of the same type of maze design. Can you find a path through it?

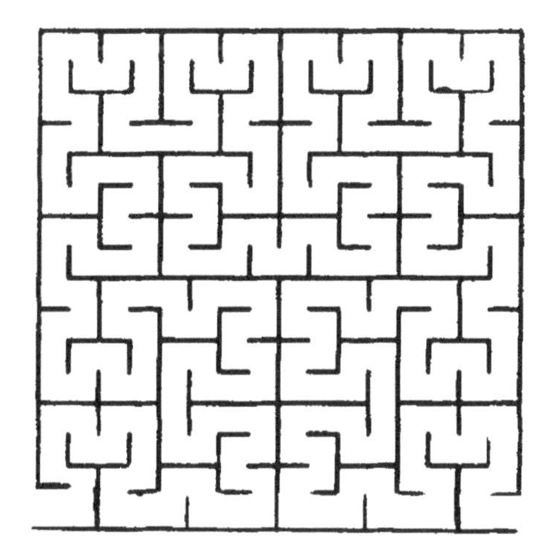

82. Now try your hand at a truly difficult maze, invented by Lewis Carroll. Can you find a path to its diamond-shaped center?

Geometry

83. As mentioned earlier, there is a much more tricky version of the bug problem, from the pen of Henry E. Dudeney. In a room 30 feet long, 12 feet wide, and 12 feet high, a spider sits in the center of one of the smaller walls, 1 foot from the ceiling; and a fly clings to the middle of the opposite wall, 1 foot from the floor. What is the shortest possible route along which the spider may crawl to reach its prey?

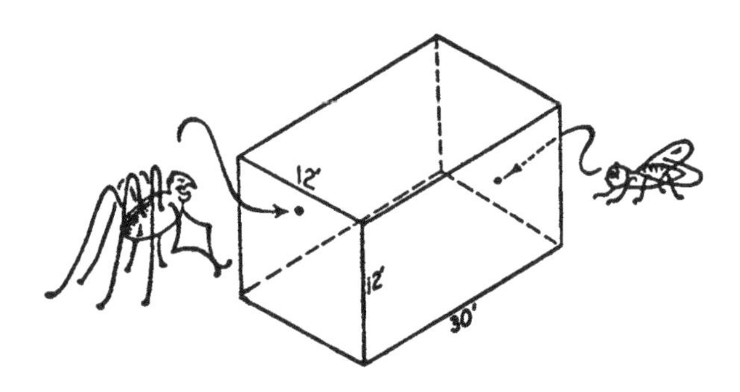

84. The first four triangular numbers are, as discussed earlier, 1, 3, 6, and 10. What is the twelfth? Do you detect any pattern?

85. The first four square numbers are, again as discussed previously, 1, 4, 9, and 16. Do you detect any pattern?

Further Reading

Boob, P. *The Idea of the Labyrinth*. Ithaca N.Y.: Cornell University Press, 1990.

Fisher, Adrian, and Georg Gerster. *The Art of the Maze*. London: Seven Dials, 2000.

Mathews, W. H. *Mazes and Labyrinths: Their History and Development*. New York: Dover, 1970.

Meehan, A. *Maze Patterns*. New York: Thames and Hudson, 1993.

ANSWERS AND EXPLANATIONS

The Riddle of the Sphinx

1.

ANSWER
Fleas

EXPLANATION
If you "catch" fleas on your body, you can, of course, "throw them away." However, if you cannot "catch" them, then you must resign yourself to "keeping" them.

2.

ANSWER
A mule

EXPLANATION
A mule is of "mingled race." It is a "half donkey" and "half horse." More specifically, it is the sterile offspring either of a male donkey and a mare or of a female donkey and a stallion. Thus, the mule is "unlike its mother" and "does not resemble its father." Because it is sterile, the mule "is incapable of producing its own progeny."

3.

ANSWER
A dog

EXPLANATION
In common parlance, a dog is said to have a "master," whose "foes" it does indeed "scare away" by bearing the "weapons" (sharp teeth or fangs) in its jaws. However, if lashed by even a child, it flees.

4.

ANSWER

A rainbow

EXPLANATION

Red, blue, purple, and green are the colors of the rainbow. Everyone can see a rainbow, but no one can ever reach it or touch it.

5.

ANSWER

Today

EXPLANATION

To grasp the answer, assume that *today* is Tuesday. Before it was born, or came to be, Tuesday did indeed have a different name—*tomorrow*. Why? Because the day before its birth was Monday. And on Monday, we would have referred to Tuesday as "tomorrow." And when Tuesday is "no more," it takes on a new name—*yesterday*. Why? Because when Tuesday ends, Wednesday comes into being. And on Wednesday, we would refer to Tuesday as "yesterday." Thus, though it lasts only one day, *today* does indeed change its name three days in a row—*yesterday*, *today*, and *tomorrow*.

6.

ANSWER

Your name

EXPLANATION

The answer is self-explanatory.

7.

ANSWERS

Only one illustrative riddle for each word is given here. Readers will undoubtedly have come up with many others of their own.

A. "I can be weighed and I am blind, but I am neither substance nor human. What am I?"

B. "I can blossom and grow, but I am neither plant nor tree. What am I?"

C. "I can be bitter or sweet, but I am neither food nor drink. What am I?"

D. "It flies but has no wings. What is it?"

EXPLANATIONS

Justice is something that we conceive as a substance that can be weighed (*the scales of justice, weighing evidence*) or as something that is blind (*justice is blind*).

We commonly say that friendship is something that blossoms and something that can *grow*, like a plant or a tree.

Love is something that we perceive as having a taste, as can be seen in expressions such as *love is sweet or love is bitter*.

The common saying *Time flies*! is the basis for the last riddle.

8.

PROOF

Start by joining vertex **A** of the triangle **ABC** to the center of the circle **O**, producing the straight line **AO**:

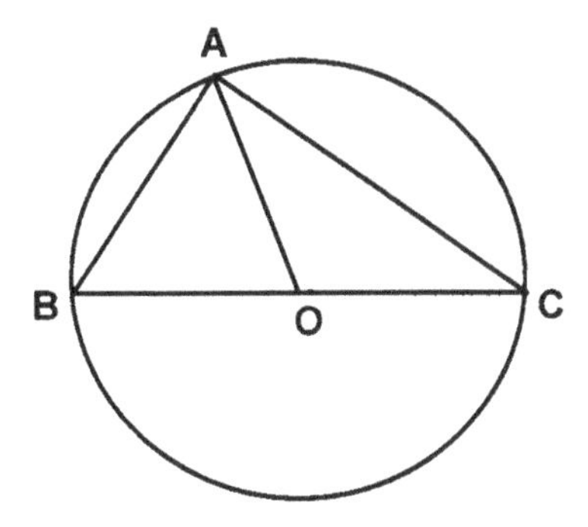

Notice that **AO** is a radius of the circle. So, too, are **OB** and **OC**. These lines are thus all equal to each other. This fact can be shown with a small stroke on each of the lines:

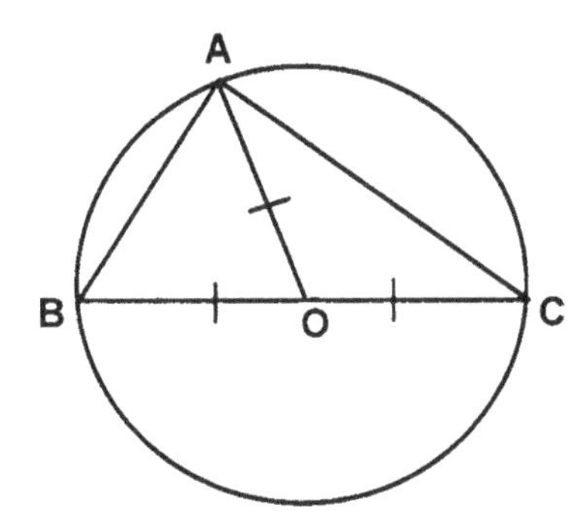

In the semicircle, there are now two isosceles triangles—**AOB** and **AOC** (as the strokes make obvious). The angles opposite the equal sides of an isosceles triangle are equal. The equal angles in triangle **AOB** can be labeled x, and those in triangle **AOC** can be labeled y:

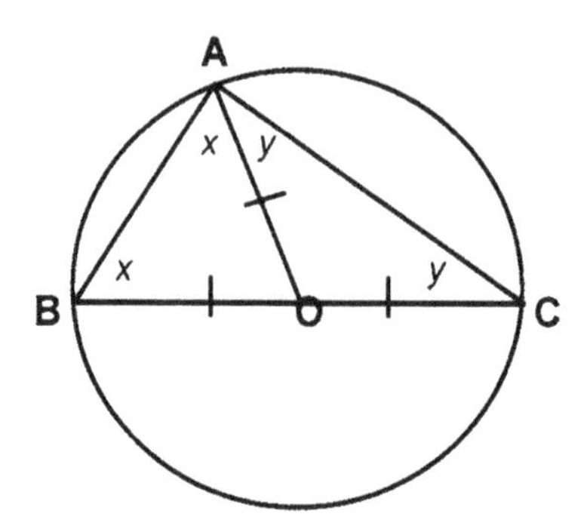

Now, consider the original triangle **ABC**. In terms of x and y, its three angles can be represented as follows:

1. $\angle$**BAC** $= (x + y)$
2. $\angle$**CBA** $= x$
3. $\angle$**BCA** $= y$

The sum of the angles in any triangle is equal to 180 degrees. Therefore, the sum of the angles in triangle **ABC** can be represented by the following equation:

$$(x + y) + x + y = 180°.$$

Simplifying the left side of the equation, we get

$$2x + 2y = 180°.$$

Dividing both sides by 2, the equation simplifies further to

$$x + y = 90°.$$

Now, $x + y$ is the total number of degrees in the angle at vertex **A**, namely, $\angle$**BAC**. Since we have just proven that $x + y$ is equal to 90 degrees, we can conclude that $\angle$**BAC** is equal to 90 degrees.

9.

PATTERN

The digits in the product of any number multiplied by 9 add up to 9 or to a multiple of 9 (18, 27, 36, etc.). The digits in any multiple of 9 also add up to 9 (or to a multiple of 9): $1 + 8 = 9$, $2 + 7 = 9$, $3 + 6 = 9$, and so on:

$9 \times 9 = 81 \rightarrow 8 + 1 = 9$

$9 \times 7 = 63 \rightarrow 6 + 3 = 9$

$9 \times 12 = 108 \rightarrow 10 + 8 = 18 \rightarrow 1 + 8 = 9$

$9 \times 100 = 900 \rightarrow 9 + 0 + 0 = 9$

$9 \times 4{,}579 = 41{,}211 \rightarrow 4 + 1 + 2 + 1 + 1 = 9$

and so on.

10.

ANSWERS

A. 477 is a multiple of 9: $4 + 7 + 7 = 18 \rightarrow 1 + 8 = 9$

B. 648 is a multiple of 9: $6 + 4 + 8 = 18 \rightarrow 1 + 8 = 9$

C. 8,765 is not a multiple of 9: $8 + 7 + 6 + 5 = 26 \rightarrow 2 + 6 = 8$ (not 9)

D. 738 is a multiple of 9: $7 + 3 + 8 = 18 \rightarrow 1 + 8 = 9$

E. 9,878 is not a multiple of 9: $9 + 8 + 7 + 8 = 32 \rightarrow 3 + 2 = 5$ (not 9)

11.

PATTERN

The square of an even number is even. The square of an odd number is odd. Thus, the square of 22, being an even number, will be even: $22^2 = 484$. And the square of 23, being an odd number, will be odd: $23^2 = 529$.

EXPLANATION

The formula for an even number is $2n$. This generalizes the fact that multiplying any number, n, by 2 will always yield an even number:

n	$2n$
0	$2 \times 0 = 0$
1	$2 \times 1 = 2$
2	$2 \times 2 = 4$
3	$2 \times 3 = 6$
4	$2 \times 4 = 8$
5	$2 \times 5 = 10$
...	

Now, let's square the formula:

$$(2n)^2 = 4n^2.$$

Is $4n^2$ also an even number? If so, we have just shown that the square of an even number is even. Notice that it can be factored as follows:

$$4n^2 = 2(2n^2).$$

The expression $2(2n^2)$ consists of $(2n^2)$ multiplied by 2. It thus represents an even number.

The formula for an odd number is $2n + 1$. This generalizes the fact that multiplying any number, n, by 2 and then adding 1 to the result will always yield an odd number:

n	$2n + 1$
0	$2 \times 0 + 1 = 1$
1	$2 \times 1 + 1 = 3$
2	$2 \times 2 + 1 = 5$
3	$2 \times 3 + 1 = 7$
4	$2 \times 4 + 1 = 9$
5	$2 \times 5 + 1 = 11$
...	

Now, let's square the formula:

$$(2n + 1)^2 = 4n^2 + 4n + 1.$$

Is the result, $4n^2 + 4n + 1$, an odd number? If so, we have just shown that the square of an odd number is odd. The expression can be written equivalently as $(4n^2 + 4n) + 1$. Now, let's factor it:

$$(4n^2 + 4n) + 1 = 2(2n^2 + 2n) + 1.$$

The result, $2(2n^2 + 2n) + 1$, represents an odd number. If you do not see this, replace $(2n^2 + 2n)$ in $2(2n^2 + 2n) + 1$ with any letter, say, m. The expression then becomes $2(m) + 1$ or $2m + 1$. This is, of course, the formula for any odd number.

For readers who may have forgotten their school algebra, the procedure for squaring $(2n + 1)$ is as follows:

$$(2n + 1)^2 = (2n + 1)\ (2n + 1).$$

Multiply the first two terms in each expression:

$$(2n + 1)\ (2n + 1) = 4n^2 + \ldots$$

Multiply the inner terms and outer terms, adding the products to the previous result:

$$(2n + 1)\ (2n + 1) = 4n^2 + 2n + 2n \ldots = 4n^2 + 4n + \ldots$$

Multiply the last two terms, adding the product to the previous result:

$$(2n + 1)\ (2n + 1) = 4n^2 + 4n + 1.$$

12.

ANSWER

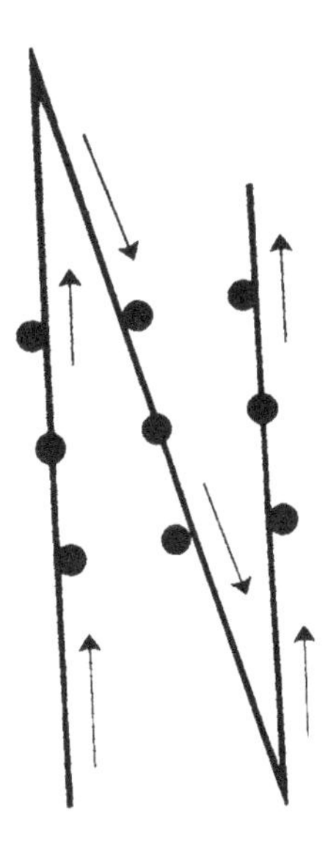

EXPLANATION

The three lines do not have to go through the center of all the dots; they can just graze some of them, as shown. That is the relevant insight required for this version of the puzzle.

13.

ANSWER

Five lines are needed to solve this version of the puzzle.

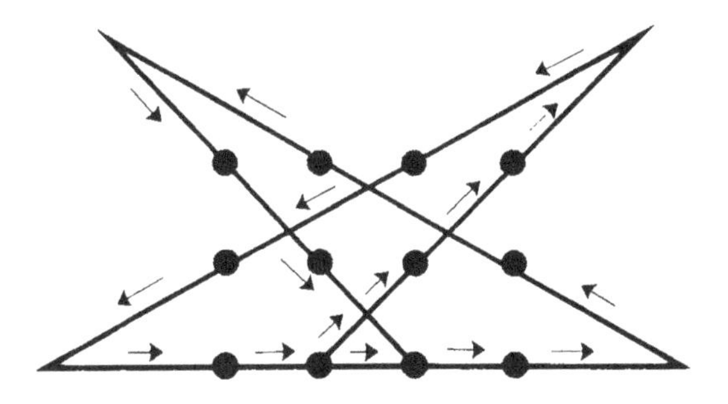

14.

ANSWER

Six lines are needed for this version of the puzzle.

Alcuin's River-Crossing Puzzle

15.

ANSWER

Five crossings are required (H_1 and W_1 = first husband-and-wife pair; H_2 and W_2 = second husband-and-wife pair):

	On the Original Side	On the Boat	On the Other Side
0.	H_1 W_1 H_2 W_2	__ __	__ __ __ __
1.	__ __ H_2 W_2	H_1 W_1 →	__ __ __ __
2.	__ __ H_2 W_2	← W_1 __	H_1 __ __ __
3.	__ W_1 __ __	H_2 W_2 →	H_1 __ __ __
4.	__ __ W_1 __	← H_1 __	__ __ H_2 W_2
5.	__ __ __ __	H_1 W_1 →	__ __ H_2 W_2
0.	__ __ __ __	__ __	H_1 W_1 H_2 W_2

16.

ANSWER

Twelve complete crossings are required. A complete crossing goes from one side to the other. A transit stop at the island, followed by a doubling back, is not considered a complete crossing. Following is one possible modeling of the four-couple version of Tartaglia's puzzle (H_1 and W_1 = first husband-and-wife pair; H_2 and W_2 = second husband-and-wife pair; H_3 and W_3 = third husband-and-wife pair; H_4 and W_4 = fourth husband-and-wife pair):

	On the Original Side	**On the Boat**	**On the Island**	**On the Boat**	**On the Other Side**
0.	H_1 W_1 H_2 W_2 H_3 W_3 H_4 W_4				
1.	H_1 H_2 H_3 W_3 H_4 W_4	W_1 W_2 →	(bypass)	W_1 W_2 →	
2.	H_1 H_2 H_3 W_3 H_4 W_4	← W_2	(bypass)	← W_2	W_1
3.	H_1 H_2 H_3 H_4 W_4	W_2 W_3 → ← W_2	(stop) W_3 (double back)		W_1
4.	H_1 H_3 H_4 W_4	H_2 W_2 →	W_3 (bypass)	H_2 W_2 →	W_1
5.	H_1 H_3 H_4 W_4	← W_1	W_3 (bypass)	← W_1	H_2 W_2
6.	H_3 H_4 W_4	H_1 W_1 →	W_3 (bypass)	H_1 W_1 →	H_2 W_2
7.	H_3 H_4 W_4	← W_3	W_1 (change)	← W_1	H_1 H_2 W_2
8.	H_4 W_4	H_3 W_3 →	W_1 (bypass)	H_3 W_3 →	H_1 H_2 W_2
9.	H_4 W_4	← W_3	W_1 (bypass)	← W_3	H_1 H_2 W_2 H_3
10.	H_4	W_3 W_4 →	W_1 (bypass)	W_3 W_4 →	H_1 H_2 W_2 H_3
11.	H_4	← W_4	W_1 (bypass)	← W_4	H_1 H_2 W_2 H_3 W_3
12.		H_4 W_4 →	W_1 (bypass)	H_4 W_4 →	H_1 H_2 W_2 H_3 W_3
			W_1	← H_1	H_2 W_2 H_3 W_3 H_4 W_4
			(double back)	H_1 W_1 →	H_2 W_2 H_3 W_3 H_4 W_4
0.					H_1 W_1 H_2 W_2 H_3 W_3 H_4 W_4

17.

ANSWER

One solution to Kirkman's puzzle is given as follows:

Monday		
0	5	10
1	6	11
2	7	12
3	8	13
4	9	14

Tuesday		
0	1	4
2	3	6
7	8	11
9	10	13
12	14	5

Wednesday		
1	2	5
3	4	7
8	9	12
10	11	14
13	0	6

Thursday		
4	5	8
6	7	10
11	12	0
13	14	2
1	3	9

Friday		
4	6	12
5	7	13
8	10	1
9	11	2
14	0	3

Saturday		
10	12	3
11	13	4
14	1	7
0	2	8
5	6	9

Sunday		
2	4	10
3	5	11
6	8	14
7	9	0
12	13	1

18.

ANSWER

The number of draws needed is three.

EXPLANATION

Suppose we draw out a white ball first. If we are lucky, the next ball we draw out will also be white, and the game is over. But we cannot assume this. We must, on the contrary, assume a "worst-case scenario"—namely, that the second ball that we will draw out is black, because the puzzle requires us to be "sure" to have a pair of balls of matching color. So, after two draws, we will have one white and one black ball. Obviously, we could have drawn out a black ball first and a white one second. The end result would have been the same: one white and one black ball.

Now, the next ball drawn from the box will, of course, be either white or black. No matter what color the *third* ball is, it will match the color of one of the two already drawn out. We will thus have a pair of balls of matching color. So, the least number of balls we will need to draw from the box in order to ensure a pair of matching balls is three.

19.

ANSWERS

A. For ten white, ten black, and ten green balls, the number of draws required is four. Assuming the worst-case scenario of drawing out one white, one black, and one green ball, in whatever order, the matching ball comes on the fourth draw because it will be a color of one of the three previously drawn balls.

B. For ten white, ten black, ten green, and ten yellow balls, the number of draws required is five. The reason is the same. After having drawn out four differently colored balls (the worst-case scenario), the fifth ball will match one of them.

C. For ten white, ten black, ten green, ten yellow, and ten red balls, the number of draws required is six. Again, the reasoning is the same. After having drawn out five differently colored balls (the worst-case scenario), the sixth ball will match one of them.

EXPLANATION

The pattern that is involved is that one more draw than the number of colors is required to ensure that a pair of balls of matching color is drawn out.

TABLE A-1: SOLUTION TO DRAWING PUZZLE

Number of Colors in the Box	Number of Draws Required to Obtain a Pair of Balls of Matching Color	Pattern
2	3	One more draw than the number of colors
3	4	One more draw than the number of colors
4	5	One more draw than the number of colors
5	6	One more draw than the number of colors
. . .	. . .	. . .
n	$n + 1$	One more draw than the number of colors

20.

ANSWER

It certainly matters if the balls varied.

EXPLANATION

When the number of balls is the same (for example, ten white and ten black, or five white and five black, and so on), then each ball has the same probability of being drawn. However, if one increases the number of balls in a specific color to, say, fifteen black, while keeping the number of balls in the other colors the same, then the chances of drawing out a black ball increase on each draw. Determining the probability of each ball would now enter into the solution, changing the nature of the puzzle.

21.

ANSWER

The least number of draws you would have to make is thirteen.

EXPLANATION

There is a total of twenty-four gloves in the box.

six pairs of black gloves = twelve black gloves

six pairs of white gloves = twelve white gloves

Of the twenty-four, half are right-handed and half are left-handed. In a worst-case scenario, we might pick all twelve left-handed gloves (six of which are black and six white) or all twelve right-handed gloves (six of which are black and six white). The thirteenth glove drawn will match one of the previous twelve.

Assume that we have drawn out the twelve left-handed gloves—six black and six white. The thirteenth draw can produce only a right-handed glove because there are no more left-handed gloves left in the box. And it will be black or white. In either case, it will be a matching color.

22.

ANSWER

The chances are two out of three.

EXPLANATION

Let B stand for the black counter that might be in the bag at the start and W_1 for the white counter that might be in the bag at the start. Let W_2 stand for the white counter added to the bag.

Let's assume first that a white counter, W_1, was in the bag at the start. When the white counter W_2 is added to the bag, it will contain two white counters—W_1 and W_2. So, the white counter that is drawn out is either W_1, the original counter in the bag, or W_2, the counter that was added to it.

Let's assume instead that the black counter, B, was in the bag at the start. When the white counter W_2 is added to the bag, it will contain a black and a white counter—B and W_2. So, the white counter that is drawn out is the one that was added to it, W_2.

Let's summarize the three scenarios in chart form:

	Inside the Bag	**Counter Added**	**Counter Drawn Out**
Scenario 1:	W_1	W_1 W_2	W_1
Scenario 2:	W_1	W_1 W_2	W_2
Scenario 3:	B	B W_2	W_2

In scenarios 1 and 2, only a white ball can be drawn. However, in scenario 3 either a white or a black ball could have been drawn (even though a white one was actually the one that was drawn). Thus, of the three scenarios, two assure us that a white ball will be drawn. The chances of drawing a white ball are, thus, two out of three.

23.

ANSWER

There are twelve routes.

EXPLANATION

There are three different ways to go from Sarah's to Bill's house. Once we are at Bill's house, there are four different ways to get to Shirley's house. So, for each route taken from Sarah's house to get to Bill's house, four routes can be taken to get to Shirley's house. There are thus $3 \times 4 = 12$ different ways altogether to get to Shirley's house from Sarah's house.

The routes can be shown in schematic form as follows. First, we represent the three possible routes from Sarah's house to Bill's house as B_1, B_2, and B_3 and the four routes from Bill's house to Shirley's house as S_1, S_2, S_3, and S_4. The twelve possible routes to Shirley's house are as follows:

From B_1 to Shirley's House

$B_1 - S_1$

$B_1 - S_2$

$B_1 - S_3$

$B_1 - S_4$

From B_2 to Shirley's House

$B_2 - S_1$

$B_2 - S_2$

$B_2 - S_3$

$B_2 - S_4$

From B_3 to Shirley's House

$B_3 - S_1$

$B_3 - S_2$

$B_3 - S_3$

$B_3 - S_4$

24.

ANSWER

There are 380 possible outcomes to the election. If only two specific members are to be elected for the presidency, there are only thirty-eight outcomes to the election.

EXPLANATION

Once a president is chosen, there are nineteen members left who can be chosen as vice president. So, there are $20 \times 19 = 380$ possible outcomes to the election. This is an example of the permutation of n objects taken r at a time: $n!/(n-r)!$ In this case, $n = 20$ and $r = 2$:

$$n!/(n-r)! = 20!/(20-2)! = 20!/18! = 20 \times 19 = 380.$$

If only Brenda and Heather can be selected for the presidency, the choice for that position is restricted to: $2! = 2 \times 1 = 2$. Now, for each one of these, there are nineteen members left (including either Brenda or Heather) for the vice president position. So, in this case, the number of possible election pairs is $2 \times 19 = 38$.

25.

ANSWER

Alex can make 792 different kinds of soup.

EXPLANATION

If he used all twelve vegetables, he could, of course, produce 12! kinds of soup. However, he is restricting the number to five. So, in total, he has $12 \times 11 \times 10 \times 9 \times 8 = 95{,}040$ possible choices. The order in which these are chosen, moreover, is irrelevant. How many redundant choices are there among these? There are $5! = 5 \times 4 \times 3 \times 2 \times 1 = 120$ of them. So, he can make $95{,}040 \div 120 = 792$ different kinds of soup.

Fibonacci's Rabbit Puzzle

26.

ANSWER

It would be a gargantuan task to list all the patterns that have so far been documented. Here is one more:

> Starting with 2, the number after every sixty numbers ends in 1: for instance, the sixtieth number after 2 is 4,052,739,537,881 (which ends in 1); the sixtieth number after that one is 14,028,366,653,498,915, 298,923,761 (which also ends in 1); and so on.

Readers who are interested in more patterns can consult the sources provided in Further Readings.

27.

ANSWER

Among several others, I have found the following interesting patterns, which hold as far as I had the patience to check:

- Starting at 6, the ratio of two consecutive terms appears to be relatively constant at 0.54 (6/11 = 0.545, 11/20 = 0.55, 20/37 = 0.540, 37/68 = 0.544, 68/125 = 0.544, etc.)
- Every number in an even position in the sequence is itself even: for instance, 2 is in the second position, 6 in the fourth, 20 in the sixth, and so on.
- Every number in an odd position in the sequence is itself odd: for instance, 1 is in the first position, 3 in the third, 11 in the fifth, and so on.

28.

ANSWER

40 days

EXPLANATION

Since Tim smokes only two-thirds of a cigarette, he leaves a butt equal to one-third of a cigarette. This means that he can piece together a new cigarette from every three cigarettes smoked. He has twenty-seven cigarettes. With these, he produced twenty-seven butts. From these butts, Tim was able to make nine new cigarettes: 27 ÷ 3 = 9. Now, with the "new" cigarettes, he produced nine more butts. From these butts, he was able to piece together three other cigarettes: 9 ÷ 3 = 3. Finally, with those bonus three cigarettes, he produced three more butts. From these butts, he could make one last cigarette. In total, therefore, he smoked 27 + 9 + 3 + 1 = 40 cigarettes. Since Tim smoked one cigarette per day, forty days went by before he quit his bad habit.

29.

ANSWER

There were fifty-nine people at the party.

EXPLANATION

Counting things (people, objects, letters, etc.) by ones, twos, threes, and so on, is equivalent to dividing them into sets—that is, into units, pairs, triplets, and so on. For example, if we count the twenty-six letters of the alphabet by twos, we are actually dividing the letters in half: 26 ÷ 2 = 13. There are thus thirteen pairs of letters in the alphabet, with none left over. If we count them by threes, we are actually dividing them in thirds: 26 ÷ 3. In this case, the answer is eight letter triples, with two left over. "Left over" means, of course, that two is the "remainder" left when 26 is divided by 3.

This insight opens up the solution to our puzzle. First, we divide the numbers between 50 and 60 by 3, identifying the numbers that leave a remainder of 2. This procedure translates into arithmetic the statement that if the people were counted "three at a time, there would be two left over":

50 ÷ 3 = 16, remainder = 2

51 ÷ 3 = 17, remainder = 0

52 ÷ 3 = 17, remainder = 1

53 ÷ 3 = 17, remainder = 2

54 ÷ 3 – 18, remaindcr = 0

55 ÷ 3 = 18, remainder = 1

56 ÷ 3 = 18, remainder = 2

57 ÷ 3 = 19, remainder = 0

58 ÷ 3 = 19, remainder = 1

59 ÷ 3 = 19, remainder = 2

60 ÷ 3 = 20, remainder = 0

With this procedure, we have identified the numbers 50, 53, 56, and 59 as those that leave a remainder of 2. Now, we need to find the number among these four that leaves as well a remainder of 4 when divided by 5. This procedure translates into arithmetic the statement that if the people were counted "five at a time, there would be four left over":

50 ÷ 5 = 10, remainder = 0

53 ÷ 5 = 10, remainder = 3

56 ÷ 5 = 11, remainder = 1

59 ÷ 5 = 11, remainder = 4

As readers can see, that number is 59. In sum, the number 59 is the only one between 50 and 60 that meets the two arithmetical requirements of the puzzle: (1) if it is divided by 3, it leaves a remainder of 2; and (2) if divided by 5, it leaves a remainder of 4.

30.

ANSWER

It contains one-third wine.

EXPLANATION

The puzzle tells us that container B is twice the size of A. So, let's proceed to draw the two containers, making B twice the size of A:

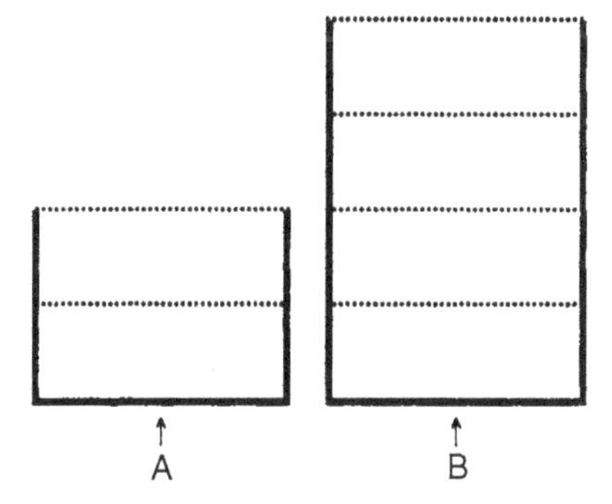

We are told that A is half filled with wine and that B is one-quarter filled with wine. With wine in them, therefore, the containers look like this:

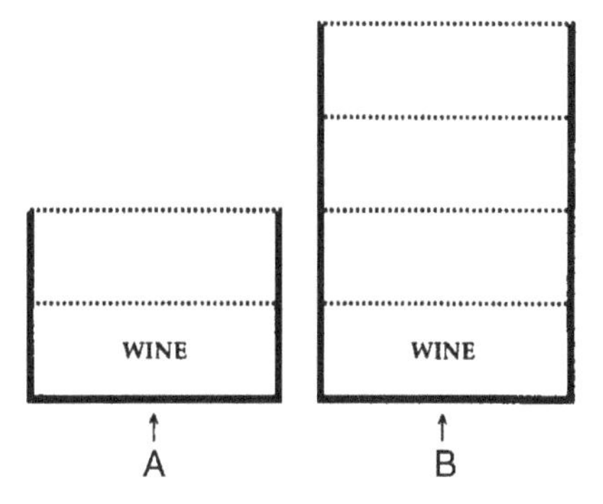

Notice that in actual fact, the same amount of wine is in the two containers. Why? Because if we calibrate the two containers into equal parts, A will have two parts and B will have four parts. The parts are all equal, because B is twice A—any one of the four parts in B is the same as any one of the two parts in A.

Now let's fill the remaining parts of the two containers with water, but without showing them "mixed up" into a solution. In reality, this is not a correct representation of what happens. It is only a convenient one:

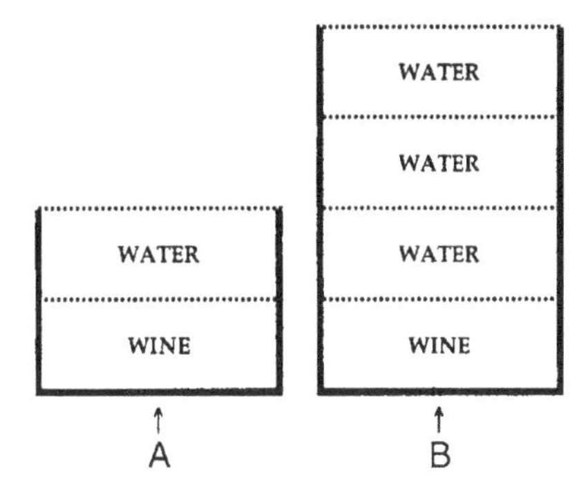

As can now be seen, A has two equal portions of wine and water, and B has three equal portions of water and one of wine. As just argued, all portions in the two containers are equal. So, between the two containers, there are six equal parts in total—two of which are wine and four water. Logically, a mixture of these two containers will contain two parts wine and four parts water. That is, in fact, what container C will have in it:

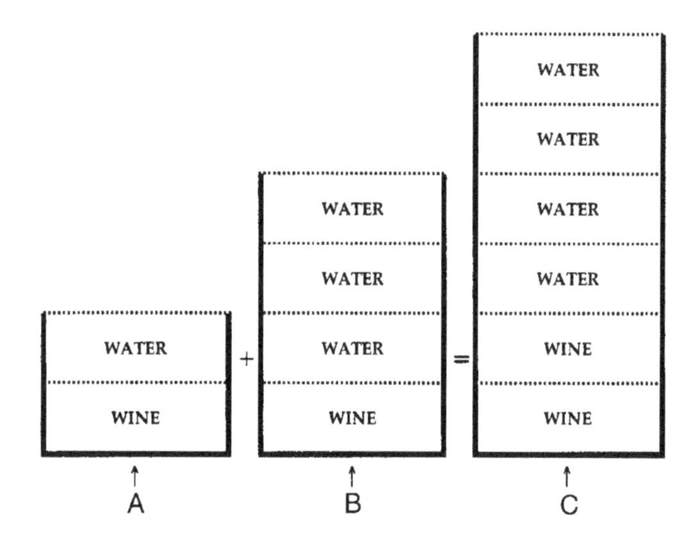

The wine and the water in container C will, of course, be blended, not separated neatly, as was shown in the previous diagram. But in that solution, wine will make up two parts out of its six, or $\frac{2}{6}$; and water will make up four parts out of its six, or $\frac{4}{6}$. In conclusion, C's mixture will have $\frac{2}{6} = \frac{1}{3}$ wine in it.

31.

ANSWER

The ladder has twenty-five rungs.

EXPLANATION

At the start, we do not know what rung the firefighter is on, except that it is the middle one. So, let's draw a model of the ladder, labeling her starting rung "0," as if it were the zero point on a number line. Each rung above and below "0" can then be compared to a digit above or below the "0" point. Obviously, since "0" is the middle rung, there will be as many rungs above it as there are below it.

We are first told that the firefighter went up three rungs from the "0" rung:

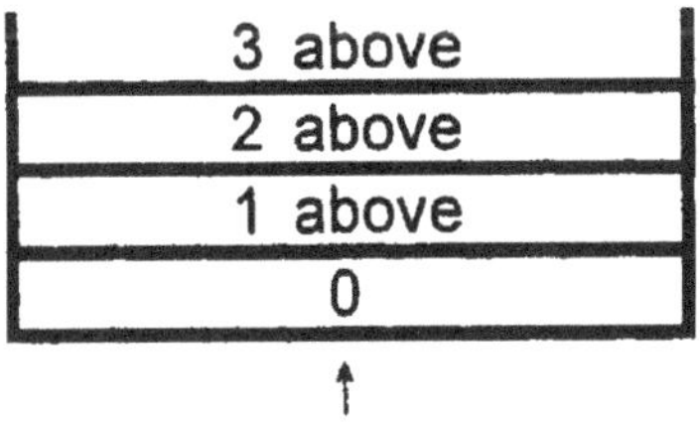

3 above
2 above
1 above
0

↑

We are then told that she stepped down five rungs. So, from rung 3 above "0," she went down five rungs, ending up at rung 2 below the starting point:

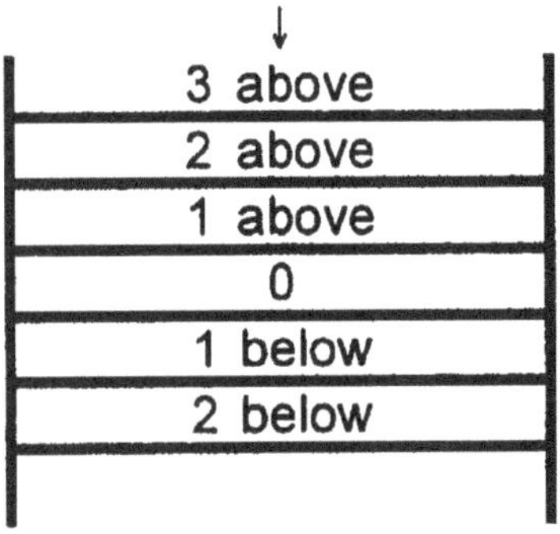

↓

3 above
2 above
1 above
0
1 below
2 below

Next, the puzzle tells us that the firefighter climbed up seven rungs. Starting from rung 2 below "0" and climbing up seven rungs from there means that she will end up at rung 5 above the middle rung:

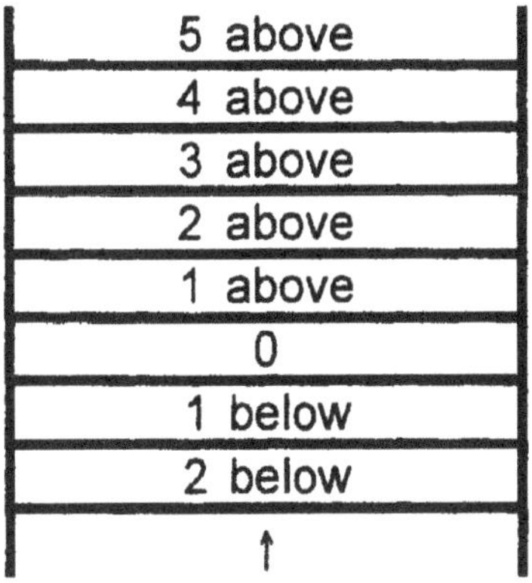

5 above
4 above
3 above
2 above
1 above
0
1 below
2 below
↑

Finally, the puzzle tells us that the firefighter climbed up another seven rungs to the roof. So from rung 5 above "0," she climbed up another seven rungs to rung 12.

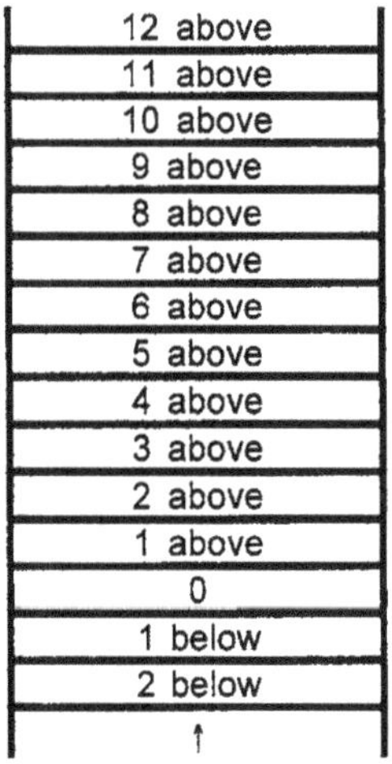

12 above
11 above
10 above
9 above
8 above
7 above
6 above
5 above
4 above
3 above
2 above
1 above
0
1 below
2 below
↑

Rung 12 is thus the last rung of the ladder. Now, let's complete the ladder. We know that it has twelve rungs above the "0" rung. Since the "0" rung is the middle rung, the ladder will also have twelve rungs below the "0" rung:

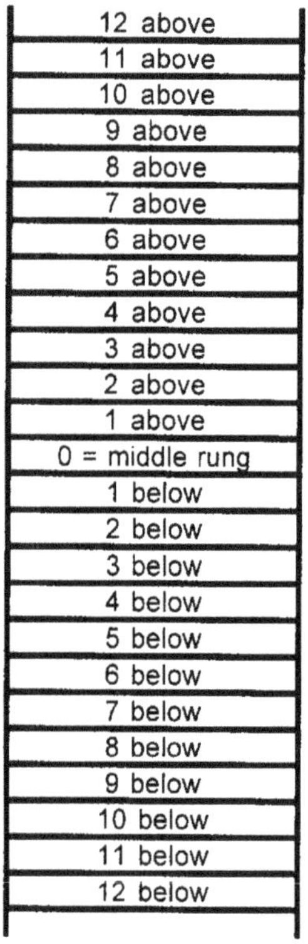

12 above
11 above
10 above
9 above
8 above
7 above
6 above
5 above
4 above
3 above
2 above
1 above
0 = middle rung
1 below
2 below
3 below
4 below
5 below
6 below
7 below
8 below
9 below
10 below
11 below
12 below

As readers can see, the ladder has twelve rungs above the "0" rung, twelve below it, and the "0" rung itself. This makes twenty-five rungs in total.

32.

ANSWER

$2\,(2^{n-1})$

EXPLANATION

Since they are, in effect, powers of 2, let's rewrite the terms of the series as follows:

{2, 4, 8, 16, 32, 64, 128, . . . }

First term: $2 = 2\,(2^0)$

Second term: $4 = 2\,(2^1)$

Third term: $8 = 2\,(2^2)$

Fourth term: $16 = 2\,(2^3)$

. . .

nth term: ?

Note that the first term is repeated in each of the successive terms, while the ratio increases by successive powers of 2. Let's rewrite the exponent in the ratio in terms of the number of the term, n. Notice in the previous example that the exponent of a particular term is one less than the number (or the position) of the term. We can now rewrite the terms of our series as follows:

First term: $2 = 2\,(2^0) = 2\,(2^{1-1})$

Second term: $4 = 2\,(2^1) = 2\,(2^{2-1})$

Third term: $8 = 2\,(2^2) = 2\,(2^{3-1})$

Fourth term: $16 = 2\,(2^3) = 2\,(2^{4-1})$

. . .

nth term: $= 2\,(2^{n-1})$

Since this pattern defines all geometric series, we can also derive a general formula for the nth term of any geometric series. We do this simply by representing the first term with the letter a and the ratio with r:

nth term

$$\begin{array}{cc} 2 & (2^{n-1}) \\ \downarrow & \downarrow \\ a & (r^{n-1}). \end{array}$$

The general term of a geometric series is ar^{n-1}.

33.

ANSWER

The sum of all the even numbers in the first 100 numbers is 2,550. The answer is not half of 5,050, or 2,525, as some readers may have assumed.

EXPLANATION

In this case, the series is

{2, 4, 6, 8, 10, 12, 14, . . . , 100}.

Line up the series with its reversed form—{100, 98, 96, 94, 92, 90, 88, . . . , 2}—under it, and then add the columns:

(1)	2	4	6	. . .	100
+	+	+	+		+
(2)	100	98	96	. . .	2
Sum	102	102	102	. . .	102.

The relevant question now becomes: how many even numbers are there in the first 100? There are, of course, 50 even numbers—half of 100. So, the sum of the numbers in the series is 50 times 102 divided by 2:

Sum: = (50) (102)/2 = 2,550.

To get the sum of the odd numbers, one could, of course, use this method again. However, since we know that the sum of all the numbers from 1 to 100 is 5,050 and that the sum of all the even numbers in it is 2,550, to compute the sum of the odd numbers, all we have to do is subtract the even number sum from the sum of all the numbers:

Sum of all the numbers		Sum of the even numbers		Sum of the odd numbers
5,050	–	2,550	=	2,500.

Euler's Königsberg Bridges Puzzle

34.

ANSWERS

Eulerian paths other than the ones given here are possible:

A. **A-F-E-B-A-D-F-C-B-D-E-C-A**

B. **A-B-C-E-B-D-E-H-D-G-H-I-E-F-I-J-F-C-A**

35.

ANSWERS

A. The graph is Eulerian because it has only two odd vertices, **F** and **H**, at which three paths converge. One or the other can be a starting or an ending vertex. Here is one possible Eulerian path, with **H** as a starting vertex: **H-G-D-E-H-I-F-E-B-D-A-B-C-F**.

B. The graph is not Eulerian because it has more than two odd vertices: **B** (= 3), **D** (= 3), **H** (= 3), **F** (= 3).

C. The graph is not Eulerian because it, too, has more than two odd vertices: **B** (= 5), **D** (= 5), **H** (= 5), **F** (= 5).

D. The graph is Eulerian because it has only two odd vertices, **F** and **G**, at which five paths converge. One or the other can be a starting or an ending vertex. Here is one possible Eulerian path: **F-C-A-F-H-G-F-D-G-E-B-D-A-B-G**.

36.

ANSWER

There are many possibilities. For a graph to be Eulerian, it can have at most two odd vertices.

37.

ANSWER

As can be seen, the octahedron has six vertices, twelve edges, and eight faces. Thus:

$$v - e + f = 2$$
$$6 - 12 + 8 = 2.$$

38.

ANSWER

A. A triangle has three edges (sides), three vertices, and one face:

$$v - e + f = 1$$
$$3 - 3 + 1 = 1.$$

B. A square has four edges (sides), four vertices, and one face:

$$v - e + f = 1$$
$$4 - 4 + 1 = 1.$$

C. A pentagon has five edges (sides), five vertices, and one face:

$$v - e + f = 1$$
$$5 - 5 + 1 = 1.$$

D. A hexagon has six edges (sides), six vertices, and one face:

$$v - e + f = 1$$
$$6 - 6 + 1 = 1.$$

In plane figures, the number of vertices equals the number of edges. And all such figures have only one face.

39.

ANSWER

One possible way to transform it into a Eulerian graph is to draw two edges externally, as shown in the following figure. These will make both **A** and **C** even (with four edges converging at each one) and thus transform a non-Eulerian graph into a Eulerian one, since it now has only two odd vertices (**B** and **D**):

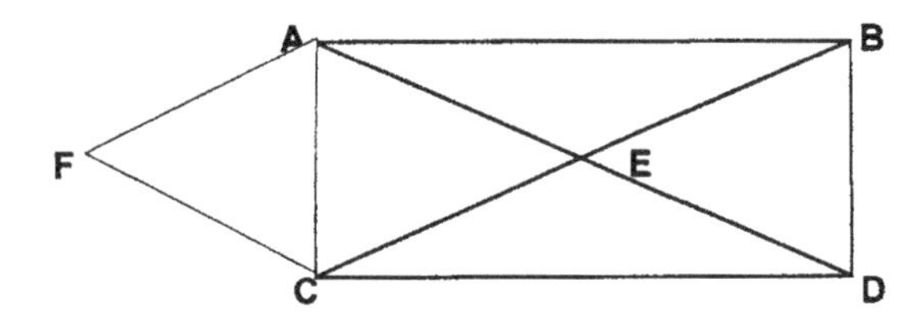

One Eulerian path through it is the following: **B-A-F-C-E-B-D-E-A-C-D**

40.

ANSWER

Only when there is an even number of inversions can the puzzle be solved. Readers must thus ensure that any arrangement they make of the blocks is such that the sum of the inversions is even.

41.

ANSWER

It is impossible for the product of two consecutive odd numbers to be an even number, such as 316.

EXPLANATION

The reason for this can be shown by multiplying the formulae for two consecutive odd integers. As discussed, the formula for an odd number is $(2n + 1)$. So the next odd number is $(2n + 3)$. Multiply these two:

$$(2n + 1)\,(2n + 3) = 4n^2 + 8n + 3.$$

Group the terms of the product as follows:

$$(4n^2 + 8n) + 3.$$

Factor 2 out:

$$2(2n^2 + 4n) + 3.$$

The term $2(2n^2 + 4n)$ represents an even number, because any number multiplied by 2 will be even. When 3 is added to it, therefore, an odd digit will result.

Guthrie's Four-Color Problem

42.

ANSWERS
(Other color arrangements are possible.)

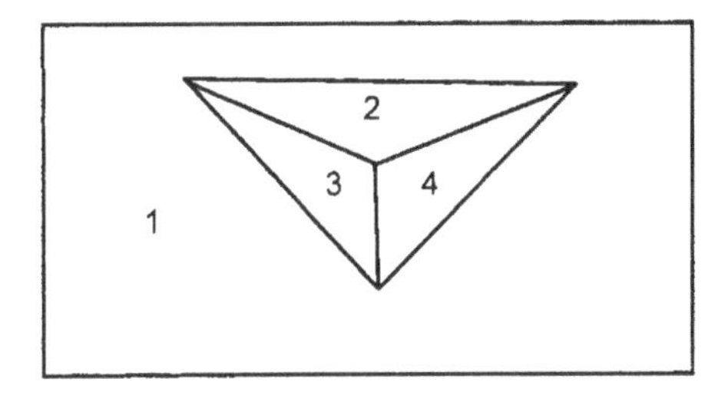

A. Four colors

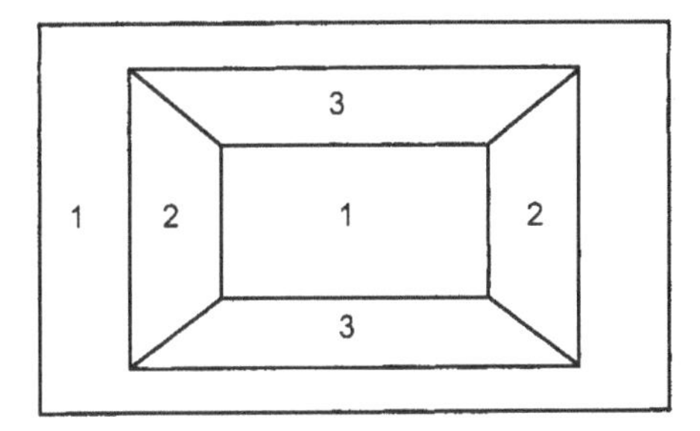

B. Three colors

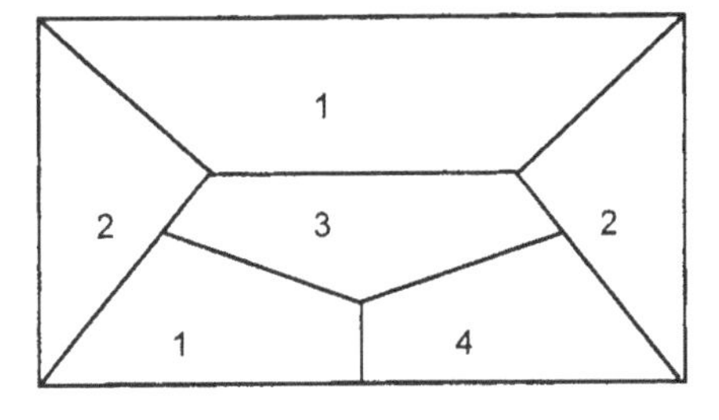

C. Four colors

1	2	3	1
3	1	2	3
1	2	3	1
2	3	1	2
3	1	2	3

D. Three colors

43.

ANSWER

Recall that both the Möbius strip and the Klein bottle have one continuing surface. Thus, only one color is required in each case.

44.

PROOF

Recall that the angles of a triangle add up to 180 degrees. If two angles in a triangle are greater than 90 degrees, then when we add them together, even without the third one, we will get a sum that is larger than 180 degrees. This goes contrary to the fact that the angles of a triangle add up to 180 degrees. Therefore, only one of the angles of a triangle can be greater than 90 degrees.

45.

PROOF

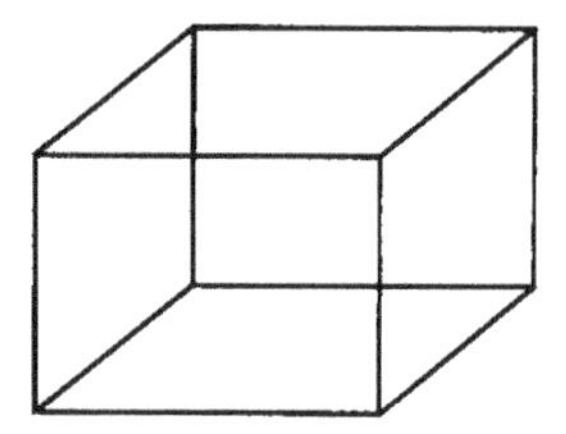

The cube has eight vertices, twelve edges and six faces. Inserting these values into the formula shows the relation to hold true:

$$v - e + f = 2$$
$$8 - 12 + 6 = 2.$$

In order to prove this, we would have to argue that it holds for all cubes. In effect, we have already done this. Why? Because the very definition of a cube is one that has eight vertices, twelve edges, and six faces. So, we were dealing with the general case from the outset. There is no exception; otherwise, the figure would not be a cube.

46.

PROOF

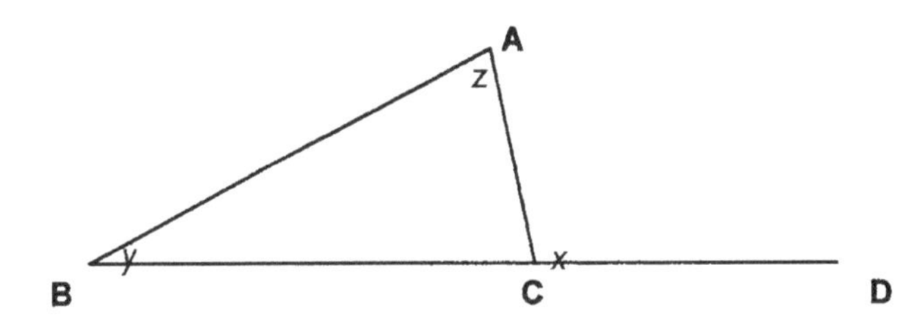

In order to prove that $x = y + z$, you can use already-established facts (theorems, propositions, etc.). Recall that the angles on the opposite side of a transversal are equal. With that knowledge, draw a line parallel to the base through vertex **A**:

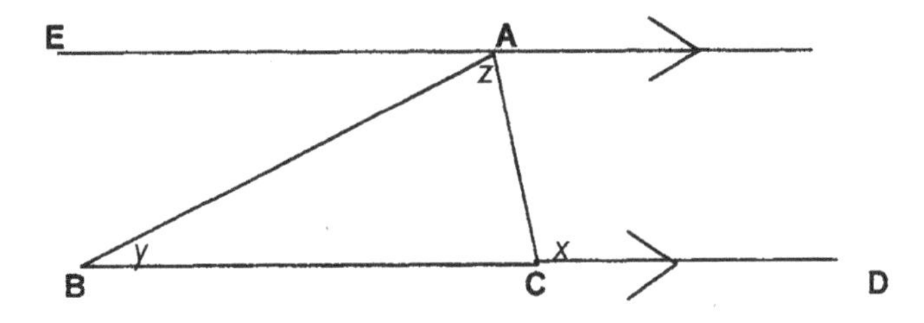

Now, we have made ∠**EAB** equal to angle y, because, as mentioned, angles on the opposite side of a transversal (**AB**, in this case) are equal. Label the angle appropriately:

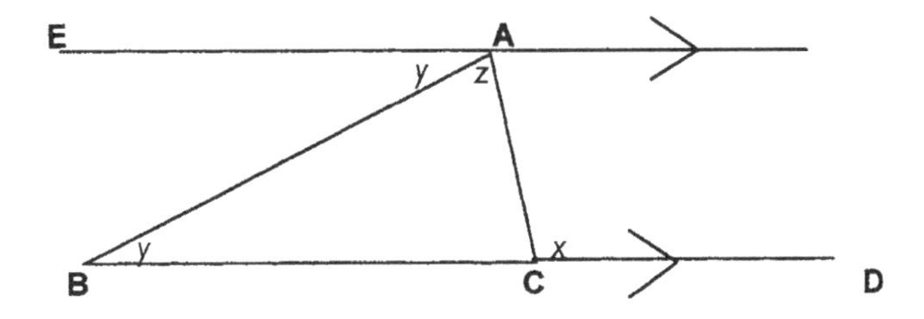

Now look at the other transversal line in the diagram, **AC**. This has made angle **EAC** equal to x. Since **EAC** $= y + z$, we have just proved that $x = y + z$ and thus that the exterior angle of a triangle is equal to the sum of the internal opposite angles.

47.

ANSWER

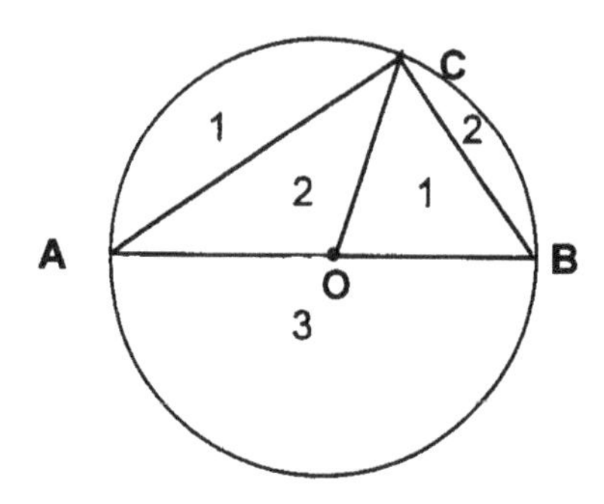

Three colors are needed.

PROOF

Notice that there are four vertices (**A-O-B-C**), eight edges (lines **AO**, **AC**, **OC**, **OB**, **CB** and arcs **AC**, **CB**, **AB**), and five faces (1, 1, 2, 2, 3) in this graph. Thus:

$$v - e + f = 1$$
$$4 - 8 + 5 = 1.$$

The five faces can be conceived to be "map regions" and the eight edges "map borders." Notice that the difference between the two is 3. This Eulerian analysis suggests that the difference between e and f on a map will indicate the number of colors required. Readers should check for themselves whether this is indeed a consistent finding, by using different types of graphs. If so, the finding would be more an example of a simple induction than it would a true proof.

Lucas's Towers of Hanoi Puzzle

48.

ANSWER

This is, of course, the equivalent of a four-disk version of the Tower of Hanoi game. So, it will take $(2^4 - 1)$ or 15 moves to play. Here is a run-down of the possible moves (ace = 1, two = 2, three = 3, four = 4), which readers can verify for themselves by getting four cards and actually playing the game:

Move	Space A	Space B	Space C
Start	1 2 3 4	——	——
1.	2 3 4	1	——
2.	3 4	1	2
3.	3 4	——	1 2
4.	4	3	1 2
5.	1 4	3	2
6.	1 4	2 3	——
7.	4	1 2 3	——
8.	——	1 2 3	4
9.	——	2 3	1 4
10.	2	3	1 4
11.	1 2	3	4
12.	1 2	——	3 4
13.	2	1	3 4
14.	——	1	2 3 4
15.	——	——	1 2 3 4

49.

PATTERN

The value of n is itself prime.

50.

PATTERN

Let's tabulate what will happen with the two given rules:

1. Rule 1 (R1) applies to an even square. It says that we must multiply the number of grains that are on the previous odd square by 2^n.
2. Rule 2 (R2) applies to an odd square. It says that the grains on the previous (even) square must be halved:

First square			$\rightarrow$		=	1
Second square	($n = 2$)	R1 applies	$\rightarrow$	1×2^2	=	4
Third square		R2 applies	$\rightarrow$	$4 \times \frac{1}{2}$	=	2
Fourth square	($n = 4$)	R1 applies	$\rightarrow$	2×2^4	=	32
Fifth square		R2 applies	$\rightarrow$	$32 \times \frac{1}{2}$	=	16
Sixth square	($n = 6$)	R1 applies	$\rightarrow$	16×2^6	=	1,024
Seventh square		R2 applies	$\rightarrow$	$1{,}024 \times \frac{1}{2}$	=	512
Eighth square	($n = 8$)	R1 applies	$\rightarrow$	512×2^8	=	131,072
Ninth square		R2 applies	$\rightarrow$	$131{,}072 \times \frac{1}{2}$	=	65,536
Etc.						

The products on each successive square turn out to be powers of 2:

First square	$\rightarrow$	1	=	2^0
Second square	$\rightarrow$	4	=	2^2
Third square	$\rightarrow$	2	=	2^1
Fourth square	$\rightarrow$	32	=	2^5
Fifth square	$\rightarrow$	16	=	2^4
Sixth square	$\rightarrow$	1,024	=	2^{10}
Seventh square	$\rightarrow$	512	=	2^9
Eighth square	$\rightarrow$	131,072	=	2^{17}
Ninth square	$\rightarrow$	65,536	=	2^{16}
Etc.				

The exponent on each odd square is one less than the exponent on the previous even square:

Second square	→	2^2
Third square	→	2^1
Fourth square	→	2^5
Fifth square	→	2^4
Sixth square	→	2^{10}
Seventh square	→	2^9
Eighth square	→	2^{17}
Ninth square	→	2^{16}

Readers may have detected other patterns as well.

51.

ANSWER

It is not possible to cover the checkerboard, for the simple reason that the two squares that were removed from it are of the same color—both white. A domino placed on the checkerboard always covers a white and a black square. With two white squares removed, the board will have more black squares on it than white squares. It does not have, therefore, an equal number of black and white squares for all the dominoes to cover.

52.

ANSWER

A.

Integers	1	2	3	4	5	6	7	8	9	10	11	12	. . .
	↕	↕	↕	↕	↕	↕	↕	↕	↕	↕	↕	↕	
Multiples of 10	10	20	30	40	50	60	70	80	90	100	110	120	. . .

B.

Integers	1	2	3	4	5	6	7	8	9	10	11	12	13	. . .
	↕	↕	↕	↕	↕	↕	↕	↕	↕	↕	↕	↕	↕	
Fractions	$\frac{1}{1}$	$\frac{1}{2}$	$\frac{1}{3}$	$\frac{1}{4}$	$\frac{1}{5}$	$\frac{1}{6}$	$\frac{1}{7}$	$\frac{1}{8}$	$\frac{1}{9}$	$\frac{1}{10}$	$\frac{1}{11}$	$\frac{1}{12}$	$\frac{1}{13}$	. . .

53.

ANSWERS

A. $\aleph_0 + 1 = \aleph_0$

B. $\aleph_0 + n = \aleph_0$

C. $\aleph_0 + \aleph_0 = 2\aleph_0 = \aleph_0$

EXPLANATION

$\aleph_0$ represents the set of the cardinal numbers. If you add 1 to it, you are simply going one number farther down the number line. Indeed, no matter how many numbers, *n*, you add to the number line, you will never go past it or beyond it. You will thus always end up on the number line. Similarly, you can double the line, whatever that means in infinite terms, but in so doing, you will not go past it or beyond it. The line is infinite and will always have the same cardinality, no matter what arithmetical operation you perform on $\aleph_0$.

Loyd's Get Off the Earth Puzzle

54.

ANSWER

The idea is to design two detached parts, A and B, which, when moved, will fit together to form a rectangle. This can be achieved by cutting up the original figure in a "zig-zag" fashion, shown as follows. The length of each cut must be equal to the length of the top right flat edge (the one jutting out). In this way, A and B will be produced as interlocking parts. By sliding A up or B down, the parts will interlock to produce a rectangle:

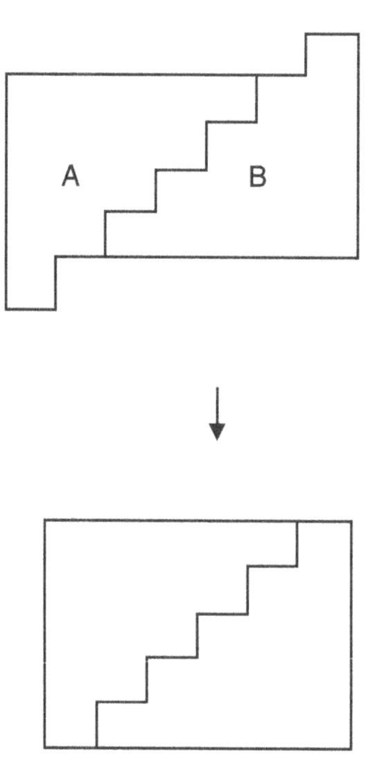

55.

ANSWER

When the two largest pieces (2 and 3) are switched, each small square that is cut by the diagonal line becomes a trifle higher than it is wide. This means that the large square is no longer a perfect square. It has increased in height by an area that is exactly equal to the area of the hole:

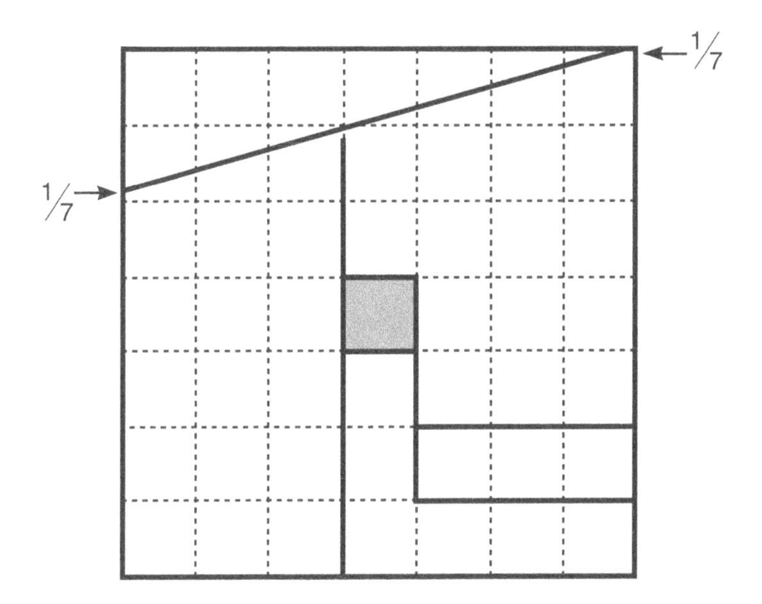

56.

ANSWER

The light- and dark-colored pencils change in number—there are seven light-colored pencils and six dark-colored ones after the reversal:

57.

ANSWER

A duck and a rabbit

58.

ANSWER

The two pencils are equal in length. This is a version of the Zöllner illusion.

59.

ANSWER

They are equal.

EXPLANATION

The radius of the largest circle is 5 units. The length of the inner radius, including everything but the outer shaded ring, is 4 units, of which 3 make up the radius of the shaded part. The area of a circle is expressed as πr^2, if you have forgotten your school geometry. Using this formula on the inner circle with a radius of 3, the area is $\pi r^2 = \pi 3^2 = 9\pi$. The area of the complete circle with a radius of 5 is $\pi r^2 = \pi 5^2 = 25\pi$. Now, to figure out the area of the outer shaded ring, we subtract this area from the area of the remainder, which has a radius of 4: $\pi r^2 = \pi 4^2 = 16\pi$. The area of the ring is thus $25\pi - 16\pi = 9\pi$. This proves that the two areas are equal, despite what our eyes tell us.

Epimenides' Liar Paradox

60.

ANSWER

The sentence leads to a circularity.

EXPLANATION

If the sentence is assumed to be true, then what it says—"This sentence is false"—must be factually true. But, if this is so, then the sentence is false (as it asserts). This would mean that the sentence is both true and false, which is logically contradictory.

So, let's assume the opposite premise—namely, that the sentence is false. What's the upshot of this new assumption? Well, if the sentence is indeed false, then the opposite of what it says must be true. But then again, this would mean that it is both false and true.

61.

ANSWER

The coin is in B.

EXPLANATION

Let's assume that the inscription on A is true:

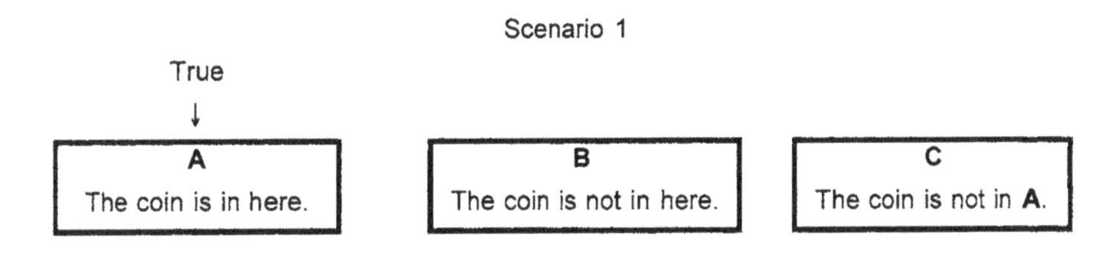

Now, we can quickly ascertain that B's inscription is also true—if the coin is in A, then, as B's inscription proclaims, it is certainly not in B. But this is contrary to the condition that at most, one inscription is true. Here we have two true statements, instead. So, we can reject scenario 1. In the process, however, we have discovered that A's inscription is necessarily false—the coin is not in A. That makes C's inscription true, since it merely confirms that the coin is not in A.

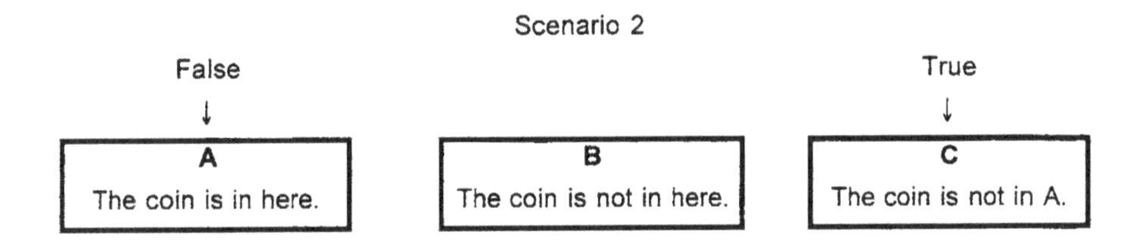

Since, at most, only one of the inscriptions is true, then B's inscription has to be false. This completes scenario 2:

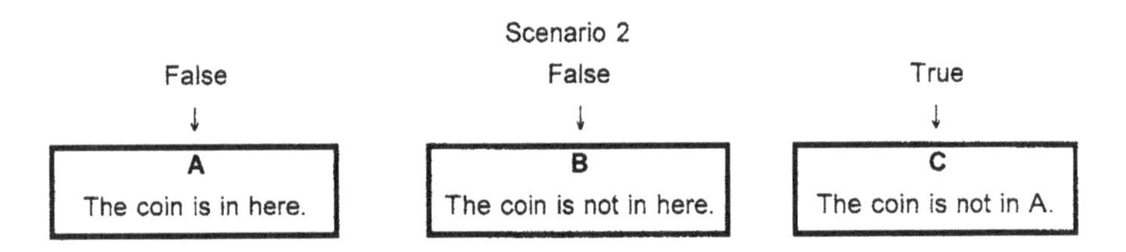

B's inscription reads: "The coin is not in here." According to scenario 2, this is a false statement. Thus the opposite is true—the coin is in B, contrary to what B's inscription says.

62.

ANSWER

It is not possible to determine who made the box.

EXPLANATION

Assume that the person who made the box was a truth-teller. Then, the inscription is false—since it says that the box was not made by a truth-teller:

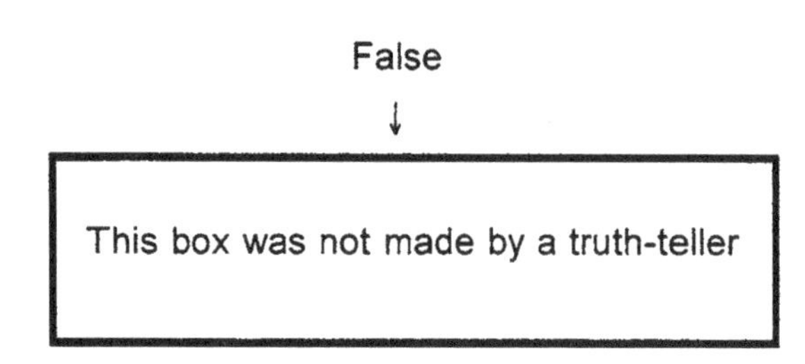

But that cannot be, because a truth-teller would not make a false inscription. So, the person who made the box must be a liar. If this is so, then the inscription is true.

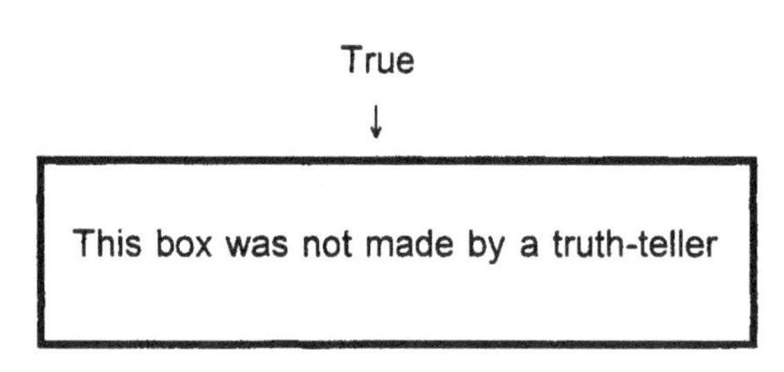

But the statement now turns out to be true, and a liar would not make such a truthful statement. So, it is not possible to determine who made the box.

63.

ANSWER

Only $x = 0$ works, because that is the only value that x can have in the equation.

EXPLANATION

Solving the equation $x + y = y$ for x, we get $x = 0$:

$$x + y = y.$$

Subtract y from both sides:

$$x + (y - y) = (y - y)$$
$$x + 0 = 0$$
$$x = 0.$$

Any other value assigned to x makes the equation impossible.

64.

ANSWER

The sentence has only five words. So, it is false. However, if we make it negative, "This sentence does not have seven words," then it does have seven words, contrary to what it says.

65.

ANSWER

The man's father

EXPLANATION

Let's call the man looking at the photo the "looker." The looker is an only child, since he has no brothers or sisters. Now, he tells us that the man in the photo has a son, who is the son of his own father. Since the looker is an only child, then he is the only possible "son" of his own father. That's who the man in the photo is.

66.

ANSWER

The bookstore salesclerk was out the $3 book and $7 from his pocket—$10 in total.

EXPLANATION

First, the bookstore salesclerk received nothing for the $3 book, since the counterfeit $10 bill was worth nothing. So, at this point, he was out $3. Now, consider what happened next. The bookstore salesclerk received ten genuine $1 bills from the record-store salesclerk, who got the counterfeit bill. When the bookstore salesclerk returned to his store, he gave $7 of the ten good bills to the customer and put the remaining good $3 in his pocket.

When the record-store salesclerk asked for her $10 back, the bookstore salesclerk still had the $3 in his pocket left over from the good $10—the other $7 went to the customer. So, he gave her back her $3 and made up the $7 difference from his own pocket. In total, therefore, the bookstore salesclerk was out the $3 book and the $7 from his pocket—$10 in total.

67.

ANSWER

For the sake of clarity, let's call the three women A, B, and C. Let's assume that A is the one who figured out the color of the cross on her head. How did she do it? A looks at B and C and sees that they both have red crosses. So, she puts up her hand, as she was instructed to do. Similarly, B also sees two red crosses. So, she, too, raises her hand. C likewise sees two red crosses; and, of course, she also raises her hand. At that point, A reasons as follows:

Let me assume that I have a blue cross on my forehead. If that is so, then one of the other two women, say B, would know that she doesn't have a blue cross because otherwise C, seeing two blue crosses—mine and B's—would not have put up her hand. But she has. So, B and C cannot determine their color. This means that I, too, have a red cross.

68.

ANSWER

The women paid $27 dollars, of which the hotel got $25 and the bellhop $2.

EXPLANATION

Originally, the women paid out $30 for the room. That's how much money was in the hands of the hotel manager when he realized that he had overcharged them. He kept $25 of the $30 and gave $5 to the bellhop to return to the women.

Now, let's focus our attention on the women. They each got back $1. This means, in effect, that they had paid $9 each for the room. Thus, altogether they paid out $27, which is $2 more than they should have paid for the room—namely, $25. As we know, these $2 were the ones pilfered by our devious bellhop!

In sum, there is no missing dollar. The women paid $27, of which the hotel got $25 and the bellhop $2.

69.

ANSWER

$$\lim_{n\to\infty} F_n / F_{n+1} = 0.618\ldots$$

Recall that this is the Golden Ratio (chapter 3).

EXPLANATION

Here are the ratios of a few consecutive Fibonacci pairs, in increasing order:

$1/2 = 0.5$

$3/5 = 0.6$

$5/8 = 0.625\ldots$

$13/21 = 0.619\ldots$

$34/55 = 0.618\ldots$

$89/144 = 0.618\ldots$

$233/377 = 0.618\ldots$

...

The ratio approaches 0.618 . . .

Recall that if F_n is any number in the Fibonacci sequence, then the symbol F_{n+1} stands for the number right after it.

70.

ANSWER

If you put compasses at the point of intersection and measure off the length of any one of the half lines, you can draw a circle passing through the ends of the lines.

EXPLANATION

The main condition that defines a circle is, in fact, that all radii are equal. Since the lines emanating from the point of intersection are drawn equal (having been bisected), then that point is really the center of a circle with a radius of *r*.

The Lo Shu Magic Square

71.

ANSWER

16	2	12
6	10	14
8	18	4

The magic square constant is 30.

72.

ANSWER

7	6	11
12	8	4
5	10	9

The magic square constant is 24.

73.

ANSWER

2.00	0.25	1.50
0.75	1.25	1.75
1.00	2.25	0.50

The magic square constant is 3.75.

74.

ANSWER

67	1	43
13	37	61
31	73	7

The magic square constant is 111.

75.

ANSWER

$2\frac{1}{2}d$	$5d$	$1\frac{1}{2}d$
$2d$	$3d$	$4d$
$4\frac{1}{2}d$	$1d$	$3\frac{1}{2}d$

The magic square constant is 9d.

76.

ANSWER

3	71	5	23
53	11	37	1
17	13	41	31
29	7	19	47

The magic square constant is 102.

77.

ANSWER

	2	3			6	7	
9			12	13			16
17			20	21			24
	26	27			30	31	
	34	35			38	39	
41			44	45			48
49			52	53			56
	58	59			62	63	

→

64	2	3	61	60	6	7	57
9	55	54	12	13	51	50	16
17	47	46	20	21	43	42	24
40	26	27	37	36	30	31	33
32	34	35	29	28	38	39	25
41	23	22	44	45	19	18	48
49	15	14	52	53	11	10	56
8	58	59	5	4	62	63	1

EXPLANATION

Note that an order 8 magic square is to be considered as made up of four smaller order 4 squares. Thus, the diagonals drawn in this case are in each of the four quadrants. After you have done this, proceed according to the algorithm.

78.

ANSWER

1. Determine the magic constant (15).

2. Determine which eight number triplets, consisting of the first nine numbers, add up to 15, because these are the triplets to be included in the eight rows, columns, and diagonals that make up an order 3 magic square:

   $9 + 5 + 1 = 15$

   $9 + 4 + 2 = 15$

   $8 + 6 + 1 = 15$

   $8 + 5 + 2 = 15$

   $8 + 4 + 3 = 15$

   $7 + 6 + 2 = 15$

   $7 + 5 + 3 = 15$

   $6 + 5 + 4 = 15$

3. From these, identify the number that occurs the most times, for this is the one to be put in the middle cell (5).

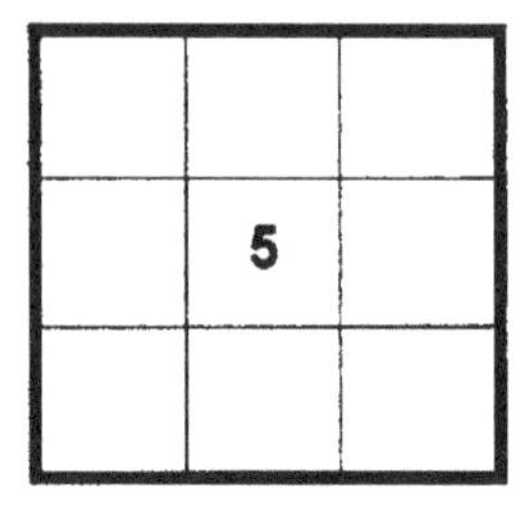

4. Now, with some trial and error, distribute the triplets, starting from the middle cell in a "radiating" pattern from that cell:

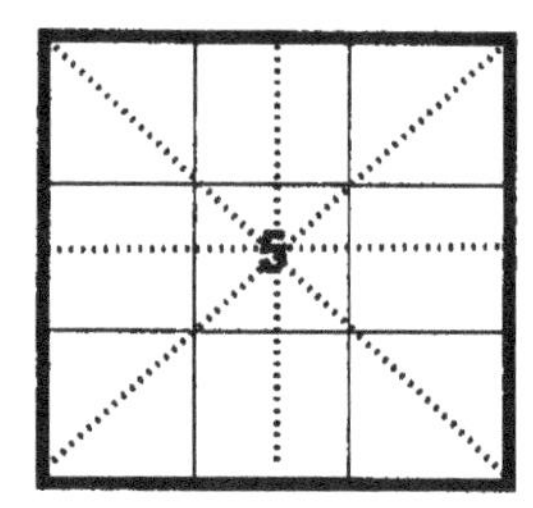

There are, of course, various ways to complete the square.

The Cretan Labyrinth

79.

ANSWER

The solution is actually quite easy. This maze has a horizontal "spoke" on the left-hand side, running from the fifth line from the outer edge to the second line from the center. If you draw a corresponding spoke on the right-hand side (from the second line from the center to the fifth line from the outer edge), the maze is reduced to a single path.

80.

ANSWER

81.

ANSWER

82.

ANSWER

83.

ANSWER

The shortest route is 40 feet.

EXPLANATION

Unfolding the room (which is a cube) will show four possible routes (A, B, C, and D). Readers can do this by making a model of the room with paper and then unfolding it in the four ways, as shown:

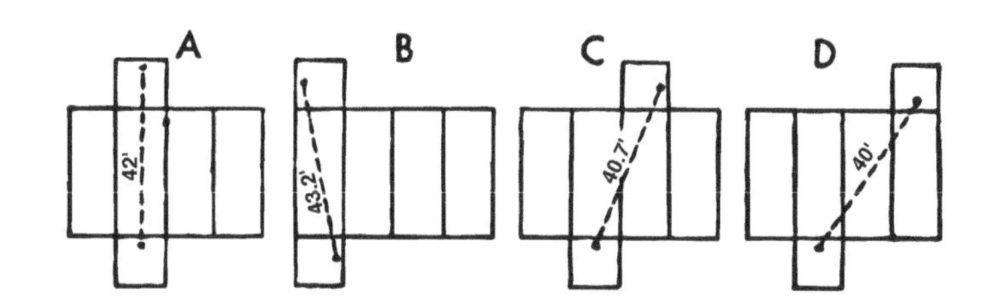

Clearly, route D is the best route. It is the hypotenuse of a right-angled triangle:

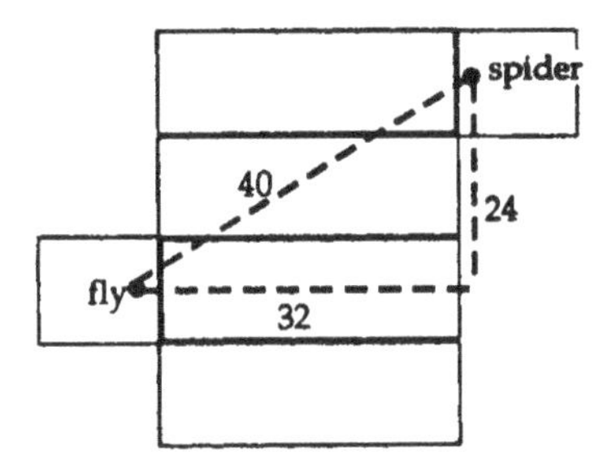

84.

ANSWER

The twelfth triangular number is 78.

EXPLANATION

Here is the pattern:

First triangular number: 1 = 1

Second triangular number: 3 = 1 + 2

Third triangular number: 6 = 1 + 2 + 3

Fourth triangular number: 10 = 1 + 2 + 3 + 4

. . .

nth triangular number: . . . = 1 + 2 + 3 + 4 + . . . n

As can be seen, each successive triangular number is produced by summing the consecutive integers in order: for example, the third triangular number is the sum of the first three integers; the seventh triangular number is the sum of the first seven integers; and so on.

So the twelfth triangular number is equal to the sum of the first twelve integers. We can now use the summation formula as follows to show that the number is 78:

$$\text{Sum}_{(n)} = \frac{n\,(n+1)}{2}$$

$$n = 12$$

$$\text{Sum}_{(12)} = \frac{12\,(12+1)}{2} = 78.$$

85.

ANSWER

The pattern is that each square number is produced by summing the consecutive odd numbers in order: for example, the third square number is the sum of the first three odd numbers; the ninth square number is the sum of the first nine odd numbers; and so on:

First square number: $1^2 = 1$

Second square number: $2^2 = 4 = 1 + 3$

Third square number: $3^2 = 9 = 1 + 3 + 5$

Fourth square number: $4^2 = 16 = 1 + 3 + 5 + 7$

. . .

Ninth square number: $9^2 = 81 = 1 + 3 + 5 + 7 + 9 + 11 + 13 + 15 + 17$

GLOSSARY

abacus A device used for showing the positional value of numerals, invented in ancient China.

acute-angled triangle A triangle in which all three angles are less than 90 degrees.

algebra A generalization of arithmetic in which symbols, usually letters of the alphabet, represent numbers or members of a specified set of numbers.

algorithm A regularized procedure for solving problems that can be used over and over again.

ambiguous figure A figure that at one time appears as something and at another as something else.

analytic geometry The study of geometric figures and properties principally by algebraic operations on variables defined in terms of position coordinates.

arithmetic The study of all types of numbers under addition, subtraction, multiplication, and division.

arithmetical series A series, such as {1, 3, 5, 7, . . . }, in which each term is formed by adding a constant to the preceding term.

axiom A self-evident or universally recognized truth ("two lines intersect at one and only one point").

base The number that is raised to various powers (the 3 in 3^2, the 4 in 4^5, and so on).

calculus The mathematical study of such concepts as the rate of change, the slope of a curve at a particular point, and the calculation of an area bounded by curves.

cardinality A term referring to the use of the cardinal numbers (positive integers) to count an infinite set.

cardinal number A number, such as 4, 15, or 948, used in counting to indicate quantity but not order.

Cartesian plane A plane having all points described by coordinates or points defined by two intersecting perpendicular lines.

chiaroscuro The technique of using light and shade in pictorial representation.

combination A grouping of elements taken from a larger set without regard to the order of the elements in each group; for example, taking two elements at a time from a set of four objects (A, B, C, and D) creates six combinations of objects: AB, AC, AD, BC, BD, CD.

combinatorics A branch of mathematics that involves the study of counting, grouping, and the arrangement of finite sets of elements.

commutativity A property of addition and multiplication by which changing the order of the numbers does not change the result: $a + b = b + a$ (e.g., $2 + 3 = 3 + 2$); $a \times b = b \times a$ (e.g., $2 \times 3 = 3 \times 2$).

composite number An integer (whole number) composed of factors: for example, $4 = 2 \times 2$.

contradiction Disproof of a proposition by showing that its inevitable conclusion leads to contradictory results or an inconsistency.

coordinate system A system of points in the plane defined by two intersecting axes.

decimal numeral A numeral constructed with ten digits (including 0), in which each digit stands for a power of ten: for example, the digit 2 stands for "two" in 2, "twenty" in 25, and "two hundred" in 250.

deduction The process of reasoning whereby a conclusion follows necessarily from the stated premises; inference by reasoning from the general to the specific; the application of general or previous knowledge to a specific problem.

dependent variable A mathematical variable whose value is determined by the value assumed by an *independent variable*.

equation A statement asserting the equality of two expressions, usually written as a linear array of symbols that is separated into left and right sides and joined by an equal sign ($x + 3 = 4$).

Eulerian path A path that traverses every edge of a graph exactly once.

exponent A small superscript number indicating the number of times that a quantity is to be multiplied by itself: $n^4 = n \times n \times n \times n$.

factor One of two or more quantities that divides a given quantity without a remainder: for example, 2 and 3 are factors of 6, because $6 = 2 \times 3$.

factorial The product of all the positive integers from 1 to a given number (written in reverse order): for example, $4! = 4 \times 3 \times 2 \times 1 = 24$.

Fibonacci number A number in the *Fibonacci sequence*.

Fibonacci sequence A sequence of numbers, {1, 1, 2, 3, 5, 8, 13, . . . } in which each successive number is equal to the sum of the two preceding numbers.

fraction An expression that indicates the quotient of two quantities: $\frac{1}{2}$, $\frac{2}{3}$, $\frac{3}{4}$, and so on.

function A variable so related to another that for each value assumed by one, there is a value determined for the other: For example, in $2x = y$, for each value assumed by x, there is one and only one value for y; y is thus a function of $2x$.

gematria An ancient craft whose basic claim was that the sum of the numerical values of letters in a name could be used to foretell such things as a person's destiny.

general case The case that refers to the whole category or to every member of a class or a category—to all points, to all angles, to all numbers, and so on.

geometric series A series, such as {1, 3, 9, 27, 81, . . . }, in which each term is multiplied by the same factor in order to obtain the next term.

geometry The study of the properties and relationships of points, lines, angles, surfaces, and solids.

Golden Ratio (Also known as the *Golden Section* and *divine proportion*); the unending number 0.6180339 . . .

graph A figure consisting of vertices, edges, and faces.

graph theory The branch of mathematics that studies *graphs*.

Hamiltonian circuit A path traced out on a graph that visits each vertex in the graph only once, except possibly for the start and finish, which may be on the same vertex.

hypotenuse The side of a *right-angled triangle* opposite the right angle.

impossible figure A figure that appears to defy common sense.

independent variable A variable whose value determines the value of other variables.

induction A process of proving that something can be established as true if it can be proved for the first and $(n + 1)$th cases; reasoning from particular facts to a general conclusion.

infinite series A series that has no final member or end value.

insight thinking A flash of insight that comes from mulling over the various aspects of a puzzle in the imagination.

integer A member of the set of positive whole numbers {1, 2, 3, . . . }, negative whole numbers {–1, –2, –3, . . . }, and zero (0).

irrational number (Also known as *radical*) any number that cannot be expressed as an integer or as a ratio between two integers, p/q ($q \neq 0$); for example, $\sqrt{2}$.

Klein bottle A one-sided surface having no inside or outside, formed by inserting the small open end of a tapered tube through the side of the tube and making it contiguous with the larger open end.

labyrinth An intricate structure of interconnecting passages or paths through which it is difficult to find one's way.

limit A number or a point that is approached by a function.

logos The faculty of mind that allows people to reason and think reflectively, considered by the ancient Greeks to be the basis of rationality and language.

Lucas sequence A sequence of numbers starting with the number 2 {2, 1, 3, 4, 7, . . . }, in which each number is the sum of the preceding two numbers.

magic square A square that contains numbers arranged in equal rows and columns in such a way that the sum of the numbers in each row, column, and diagonal is the same.

magic square constant The sum of the numbers in each row, column, and diagonal of a *magic square.*

matrix An arrangement of symbols in columns and rows.

Mersenne number A number that is produced by the formula $(2^n - 1)$.

metalanguage A language used to describe other languages.

Möbius strip A continuous one-sided surface that can be formed from a rectangular strip by rotating one end 180 degrees and attaching it to the other end.

negative number A number whose value is less than zero.

nonplanar graph A multi-dimensional graph.

obtuse-angled triangle A triangle in which one of the angles is greater than 90 degrees.

optical illusion A visually perceived image that is deceptive or misleading.

paradox An assertion that leads to a contradiction, though based on a valid deduction from acceptable premises.

Pascal's triangle A triangular arrangement of integers whereby a number in a row is the sum of two numbers immediately above it in the triangle.

perfect number A positive integer that is equal to the sum of its integral factors; for example, 6 is a perfect number because its three divisors, 1, 2, and 3 ($6 = 1 \times 2 \times 3$), add up to 6 ($6 = 1 + 2 + 3$).

permutation A grouping of elements taken from a larger set with regard to the order of the elements; for example, in making permutations of two objects from a set of four objects (A, B, C, and D), there would be four candidates to choose from for the first selection and three left over to choose from for the second selection, or twelve permutations in all.

perspective drawing The technique by which three-dimensional space can be convincingly portrayed on a two-dimensional surface.

planar graph A two-dimensional graph.

polygon A closed plane figure bounded by three or more lines (a triangle, a quadrilateral, a pentagon, a hexagon, etc.).

positive number A number whose value is greater than zero.

postulate Any statement that requires no proof since it is either self-evident or simply put forward as an hypothesis.

power A synonym for *exponent*.

primality testing A way of determining whether a number is *prime* or not with some procedure.

prime number An integer (whole number) with no factors other than 1 and itself; for example, 1, 3, 5, 7, 19, and so on.

proposition Something that is expressed in a statement.

puzzle A problem that challenges us to seek a nonobvious answer.

puzzlemath An approach to the study of fundamental mathematical ideas through the use of puzzles.

Pythagorean theorem The square on the hypotenuse (c) of a right-angled triangle is equal to the sum of the squares on the other two sides (a, b): $c^2 = a^2 + b^2$.

Pythagorean triple A set of three numbers, such as 3, 4, and 5, related to one another in terms of the *Pythagorean theorem* ($5^2 = 3^2 + 4^2$).

rational number A number that is capable of being expressed as an integer or as a quotient of integers, excluding zero as a denominator; its general form is p/q ($q \neq 0$).

reductio ad absurdum Disproof of a proposition by showing the absurdity of its inevitable conclusion.

right-angled triangle A triangle in which one of the angles is 90 degrees.

Roman numeral A numeral constructed from seven alphabet letters, each one having a specific numerical value: I = one, V = five, X = ten, L = fifty, C = one hundred, D = five hundred, M = one thousand.

root The number that is raised to various powers (the 3 in 3^2, the 4 in 4^5, and so on); also called *base*.

self-referentiality Reference of something to itself.

series A sequence of numbers, called *terms*, that is generated by some rule; for example, {2, 4, 6, 8, . . . } is a series in which each term is generated by adding 2 to the previous one.

set theory The study of the properties of sets.

square number Any number that can be represented as a square figure (1, 4, 9, 16, . . .).

syllogism A form of deductive reasoning consisting of a major premise, a minor premise, and a conclusion; for example, "All human beings are mortal" is the major premise; "I am a human being" is the minor premise; "Therefore, I am mortal" is the conclusion.

systems analysis The study of a procedure to determine the desired end and the most efficient method of obtaining this end.

theorem A statement that has been or is to be proved on the basis of explicit assumptions ("When two straight lines intersect, the vertically opposite angles that are formed are equal").

topology The branch of mathematics that studies the properties of graphs or figures that remain unaltered when they are bent, twisted, stretched, or deformed in some way.

transfinite number A number that is greater than any finite number.

transversal A line that intersects other sets of lines.

triangular number Any number that can be represented as a triangular figure (1, 3, 6, 10, . . .).

INDEX

Printed in the USA
CPSIA information can be obtained
at www.ICGtesting.com
JSHW082158140824
68134JS00014B/295